MATHEMATICS OF THE TRANSCENDENTAL

ALSO AVAILABLE FROM BLOOMSBURY

Being and Event, Alain Badiou

Conditions, Alain Badiou

Infinite Thought, Alain Badiou

Logics of Worlds, Alain Badiou

Theoretical Writings, Alain Badiou

Theory of the Subject, Alain Badiou

MATHEMATICS OF THE TRANSCENDENTAL

Alain Badiou

Edited, translated and with an introduction by A. J. Bartlett and Alex Ling

BLOOMSBURY ACADEMIC
LONDON • NEW YORK • OXFORD • NEW DELHI • SYDNEY

BLOOMSBURY ACADEMIC
Bloomsbury Publishing Plc
50 Bedford Square, London, WC1B 3DP, UK
1385 Broadway, New York, NY 10018, USA
29 Earlsfort Terrace, Dublin 2, Ireland

BLOOMSBURY, BLOOMSBURY ACADEMIC and the Diana logo
are trademarks of Bloomsbury Publishing Plc

A catalogue record for this book is available from the British Library.

Library of Congress Cataloging-in-Publication Data
Badiou, Alain.
Mathematics of the transcendental / Alain Badiou; edited, translated and with
an introduction by A. J. Bartlett and Alex Ling. pages cm
Includes bibliographical references and index.
ISBN 978-1-4411-8924-0 (hardcover)-- ISBN 978-1-4411-3038-9 (ebook (epub))--
ISBN 978-1-4411-5043-1 (ebook (pdf)) 1. Ontology. 2. Transcendental logic.
3. Categories (Mathematics) I. Bartlett, A. J. (Adam John), 1967- editor of
compilation. II. Title. BD312.B3413 2014 111--dc23 2013029986

ISBN: HB: 978-1-4411-8924-0
PB: 978-1-4742-8645-9
ePDF: 978-1-4411-5043-1
ePub: 978-1-4411-3038-9

Typeset by Fakenham Prepress Solutions, Fakenham, Norfolk NR21 8NN

To find out more about our authors and books visit
www.bloomsbury.com and sign up for our newsletters.

TABLE OF CONTENTS

TRANSLATORS' INTRODUCTION: THE CATEGORIAL IMPERATIVE

There is no difference between what I have done and what such philosophers as Plato, Descartes, Leibniz, or Hegel have done, a hundred times over since the very origins of our discipline: reorganizing a thorough, if not creative, knowledge of mathematics, by means of all the imagining powers of language.

BEING AND EVENT, XIV

I hope that I say nothing imprecise in mathematics, but also nothing that is mathematically proffered.

THEORY OF THE SUBJECT, 209–10

The two works made available here for the first time in either English or French – *Topos, or Logics of Onto-Logy: An Introduction for Philosophers* and *Being-There: Mathematics of the Transcendental* – represent Badiou's early forays into category theory (and in particular the sub-branch of topos theory) from two vastly different philosophical angles. While the former comes hot on the heels of the publication of his first 'great book', *Being and Event*, the latter captures in painstaking detail the conceptual development of what would ultimately become the belated 'sequel' to this foundational work, *Logics of Worlds*. As such, they catalogue Badiou's philosophico-mathematical trajectory over a period of nearly two decades – a period of astonishing philosophical and artistic productivity on Badiou's part[1] – from his enthusiastic embrace of set theory as the theory of pure multiplicity and consequent assertion that 'mathematics is ontology', up to his parallel claim that 'logic is appearing' together with his difficult excursions through the intricacies of topos theory.[2]

Given the intervallic nature of this work, it should come as little surprise that both books can be seen to serve equally different agendas. If in the first book Badiou approaches category theory (at least initially) as a foundational and hence *ontological* rival to set theory – noting that they each 'define entirely different directions of thought regarding the prescriptions for the foundations of philosophy'[3] – by the second this ontological competition had shifted toward something of an onto-*logical* communion, whereupon category theory came to designate the logical phenomenology that forms the complement to his set-theoretic ontology.

Simplifying brutally, Badiou's primary concern with *Topos* is that, having only recently established an entire philosophy on the basis of a specific branch of mathematics that (crucially) serves as a foundational system for mathematics itself – namely, post-Cantorian set theory – and having utilized all the conceptual machinery of this system in its elaboration, he nonetheless remained philosophically obliged to account for that other great pretender to the crown of mathematical foundations: category theory. For if mathematics is a thought (that is, if mathematics constitutes a truth-procedure) then what it thinks must itself be (re)thought by any philosophy worthy of the name *in its own terms and on its own ground.*[4]

And so, following incisive critiques made in particular by Jean-Toussaint Desanti (first at Badiou's *Habilitation* and later in a pivotal essay published in *Les Temps modernes*), Badiou quickly came to recognize that 'I could not exempt myself from proposing a philosophical interpretation of [set theory's] rival from within my system',[5] and as such urgently set about digesting the relevant literature. The work itself – which was originally drafted as a 'backup' to his 1991 doctoral seminar – no doubt bears the traces of this urgency. Remarkably concise and written with characteristic rigour and precision, *Topos* in many ways closer resembles a mathematical textbook than a philosophical treatise.[6]

By the time of *Being-There* however, Badiou's philosophical and mathematical situation had changed considerably. Far more familiar with category theory (having spent the better part of the last decade ploughing through the field's canonical works),[7] and by now firmly convinced of its relevance to his own project, Badiou's concern with this work is less to demonstrate the logico-mathematical as well as philosophical differences between the two 'rival' theories than to outline the positive philosophical application of category theory in terms of a transcendental theory of appearing or 'being-there' (which would nevertheless leave his set-theoretic ontology essentially intact).

In short, if Badiou was initially distrustful of category theory and defensive of set theory's foundational status (while at the same time clearly impressed by its technical unfolding), this scepticism transformed over time into a genuine philosophical appreciation of its logical power and its potential to provide a structure capable of accounting for the infinitely varied intensities of appearing, that is, its capacity to *formalize appearing itself*.

This intense interest in or *passion for* formalization (to echo one of the central theses of *The Century*)[8] of course in no way marks a deviation in Badiou's overall philosophico-mathematical trajectory. To the contrary, Badiou's writings have from the first been distinguished foremost by their formalizing zeal (the full title of his first book of philosophy *The Concept of Model: An Introduction to the Materialist Epistemology of Mathematics* amply testifying to this fact),[9] and his philosophy has always been to a certain degree subordinated to the mathematical imperative, not only in the form of set theory and algebra but also that of topology, analysis and number theory, along with the revolutionary theorems of Gödel, Cohen, Easton and others.[10]

Even amongst the scientifically rigorous world of the Cercle d'Épistémologie,[11] Badiou quickly set himself apart as the most mathematically engaged member of the group, publishing two decisive papers in the *Cahiers pour l'Analyse* dealing respectively with Abraham Robinson's non-standard theorization of infinitely small quantities and Gödel's incompleteness theorem.[12] Likewise, his attempts in the mid-70s to reinvigorate Marxism with a materialist and dialectical philosophy by appending to it a formal theory of the subject (in a series of lectures that would later be published as *Theory of the Subject*)[13] relied as heavily on abstract algebra, topology and the mathematics of Gödel (constructability) and Cohen (genericity and forcing) as it did on Marxian orthodoxy and the Hegelian dialectic.

If, however, mathematical formalization (in the manner of Jacques Lacan)[14] was Badiou's express objective in the mid-1960s and again later in the mid-1970s – his scientistic agenda being interrupted for a time by the events of May 1968 and the urgent need for direct political engagement – it was not until the publication of *L'Être et l'événement* in 1988 (translated into English as *Being and Event* in 2005)[15] that Badiou was able to present a truly *systematic* philosophy (even if many crucial aspects of this work are foreshadowed in *Theory of the Subject*). That this philosophy unfolded so systematically was of course due in no small part to his fundamental thesis on the nature of being and its mathematical inscription, a thesis whose manifold consequences Badiou was able to capture in a single, striking equation: *mathematics is ontology.*

It is of course in this groundbreaking work that Badiou famously holds that being itself, being *qua* being, is nothing other than *pure multiplicity*. Or again, once an object is divested of (or 'subtracted from') everything that goes into making it a 'unique' thing – once we isolate it from its context and strip away each and every one of its qualitative determinations – what remains is essentially a multiple of multiples. This pure multiple remainder is literally the *being* of the object, the elementary 'there is' underlying everything that 'is there'. Crucially, there is no 'atomic' halting point to this infinite de-composition; what we arrive at is not the 'One' (that is, some form of primordial unity), but rather the void, nothingness itself – the in-finite dissemination of multiple multiplicity.[16]

It is in direct response to this insight regarding the multiplicity of being that Badiou makes his most celebrated and wide-reaching philosophical decision, namely, that mathematics *is* ontology. For if ontology is, as Aristotle declared in his *Metaphysics*, the science of being *qua* being – and if pure being, or being subtracted from all of its particular qualities and attributes, is, as Badiou argues, none other than inconsistent (or 'uncounted') multiplicity – then the only science adequate to thinking such multiple being is mathematics. Indeed, thinking pure multiplicity is precisely what mathematics does on a daily basis. No other discourse – be it theological, linguistic, relativist, vitalist, phenomenological, or whatever – so much as comes close to being up to this task; mathematics and mathematics alone gives us the thought of being.

Needless to say, Badiou's declaration that ontology is mathematics is in no way to suggest that being is itself mathematical. To do so would be to illegitimately conflate ontology, which is simply the *discourse* on being, with the object of this discourse, namely, being itself. Rather, mathematics – or more precisely, Zermelo-Fraenkel set theory (ZFC), the axiomatic system that purports to offer a solid foundation for mathematics[17] – *is* ontology inasmuch as it provides the *minimal and sufficient* structure necessary to articulate multiple multiplicity.

In fact, ZFC literally 'thinks' pure multiplicity in a number of immediate ways. For one thing, a set in ZFC has no 'essence' other than that of being a multiple; it is determined neither externally (nothing constrains the way it seizes another thing) nor internally (a set is entirely indifferent to what it collects). Meaning a set, thought in itself, *in-consists* – it consists (or is composed) of nothing – its sole predicate being its multiple nature. To this end a set literally *is* inconsistent multiplicity. Moreover, every element of a set is itself a set, meaning every multiple is itself a multiple of multiples, without reference to any unitary element. So a set *qua* inconsistent multiplicity is radically without-One; it is, in its essence, uncounted multiple multiplicity.

The opening meditations of *Being and Event* in fact more than amply demonstrate how every ontological concept has an immediate correlate in ZFC.[18] While we have no intention of rehearsing Badiou's argument in full here – not simply for reasons of space but moreover because there already exists a surplus of summaries, examinations and exegeses of Badiou's set-theoretic ontology that are available today[19] – it is nevertheless important to highlight the fact that, while the enormously productive role of Badiou's ontology is immediately apparent (its vast edifice being entirely built out of operations which are essentially performed on *nothing*), its basic gesture is nonetheless *subtractive*. Indeed, as Badiou points out again and again, his is ultimately a 'subtractive ontology'.

This 'subtractive' operation can be understood in a number of ways. First, Badiou's ontology is subtractive insofar as it does not purport to convey being as presence. To the contrary, being – *pure being* (or pure multiple multiplicity) – is that which defies any form of presentation (or representation, for that matter). Radically withdrawn from all unification, being is nothing other than uncounted – and therefore unpresented – multiplicity. So Badiou's ontology is also subtractive in a second sense, in that it 'subtracts' being from its capture by the One (which, as Badiou argues in *Being and Event*, is the dominant trope of classical metaphysics).[20]

Perhaps most importantly however, Badiou's ontology is subtractive in that its fundamental gesture, curiously enough, is to subtract *being itself* from ontology. For ontology is ultimately a discourse which prescribes the rules by which something can be presented or 'counted' as one – its sole operation being that of the count – and the 'one' thing that necessarily *fails* to be counted is nothing other than inconsistent multiplicity, or being itself. So, technically speaking, being isn't actually given in ontology; rather, it is retroactively posited on the basis of conceiving the one as a 'result' (of the operation of the count). To repeat: pure multiplicity is not something that can be *known* (or again, while we certainly know that being *is*, we cannot know *what* it is); rather, the nature of being is something that must be *decided* upon, in an axiomatic sense.

With *Being and Event*, Badiou believed he had made his definitive philosophical statement. At the time he was, as he puts it in the introduction to his 'transitional' work, *Briefings on Existence*, 'experiencing a spirit of conquest':[21]

> Not without pride, I thought that I had inscribed my name in the history of philosophy, and in particular, in the history of those philosophical systems which are the subject of interpretations and commentaries throughout the centuries.[22]

Like Cantor before him, Badiou had entered the paradise of which Hilbert famously declared 'no one shall expel us from'. Yet doubts about the overall cohesiveness of his system quickly began to set in:

> Happy times were coming to a close. I told myself: "the idea of the event is fundamental. But the theory I propose on what the event is the name of is not clear." Or: "The ontological extension of mathematics is certain. But, then, what about logic?" Many other doubts and questions ensued.[23]

In fact, in the years following *Being and Event*, Badiou came to, if not question, then at least rearticulate almost all of his central philosophical concepts, from the nature of the event to the constitution of the subject and the truth it institutes.[24]

By far the most consequential of these post-*Being and Event* doubts – certainly the one demanding the broadest conceptual armoury – involved the question of appearing.[25] In a word, Badiou came to realize that a situation could not be understood simply as a multiple/set, but rather needed to take into account the entire network of relations this multiple sustains. Or again, inasmuch as the mathematical theory of pure multiplicity provided the thought of *presentation as such* (and no more), there needed to be some way to think everything that *is presented* in the infinity of the real situation. To this effect Badiou contended that situations must be conceived simultaneously in their *being* (i.e. as a pure multiple) as well as their *appearing*, namely, *as the effect of a transcendental legislation*.

The set-theoretic resources of *Being and Event* – resources that are constrained to recognize only absolute identity and difference[26] – were clearly incapable of providing such a relational thought; what was required was a *logic*.[27] With this 'apparent' gesture, Badiou had expelled himself from his own paradise.

As we telegraphed earlier, this 'expulsion' may never have taken place were it not for the direct intervention of Jean-Toussaint Desanti, another philosopher who, like Badiou, truly places his own thought under the mathematical condition (and thus under the condition of pure deductive rationality).[28]

Indeed, Desanti's contribution to the 1990 special issue of *Les Temps modernes* devoted to *Being and Event*, 'Quelques Remarques à propos de l'ontologie intrinsèque d'Alain Badiou'[29] (which followed on from comments he initially delivered during Badiou's *Habilitation*), proved to be something of an 'event' for Badiou's philosophy, as well as the catalyst for this book. In essence, Desanti's paper presents a sympathetic critique of what he calls Badiou's 'intrinsic ontology', which he initially commends yet ultimately finds 'too impoverished to accomplish what [Badiou] expects of it',[30] concluding by wondering whether Badiou's philosophy need restrict itself to the set-theoretic universe or if contemporary mathematics might rather 'offer possibilities that would allow for another basic ontology'.[31]

The importance of this critique cannot be overestimated. In fact, were it not for Desanti's intervention, Badiou's later philosophy would almost certainly have followed an altogether different trajectory, as it was this that led directly to Badiou's recognition that he needed to come to grips with set theory's principal rival theory regarding the possible foundations of mathematics, namely, category theory. Given the centrality of Desanti's paper to Badiou's post-*Being and Event* philosophy (and *Mathematics of the Transcendental* in particular), a brief synopsis would not be out of order here.

Beginning innocuously enough, Desanti notes that while Aristotle's famous designation of ontology as 'the science of being *qua* being' invites one to think being *intrinsically* (that is, *qua* being), this 'intrinsically' can itself be thought in two very different ways, namely, either *maximally* or *minimally*. A maximal interpretation of intrinsic ontology, Desanti observes, would ultimately involve attempting to render being adequate to its concept. That is to say, someone who adopts the maximal approach must 'try to think under the name "being" the deployment of this very concept in the richness and interconnectedness of its movements'.[32] Or again, the maximal interpretation attempts to think being 'fully', in all its infinite grandeur. A minimal approach contrarily seeks to uncover being in its being by using *the least means necessary*. That is to say, the minimal interpretation of intrinsic ontology identifies '*the least* that must be thought in order to define the status of the proposition "*there are* beings"'.[33]

We might at this point clarify some confusion that has arisen in recent scholarship regarding Desanti's 'intrinsic' ontology and Badiou's own 'subtractive ontology' (of which we spoke above). For while both the maximal and the minimal interpretations of intrinsic ontology certainly seek to think being *in itself*, or being in its being (hence its 'intrinsic' determination), only the latter is properly speaking *subtractive*. Indeed, it is immediately clear that Badiou's approach to ontology in *Being and Event* is in no way maximal but rather minimal through and through, its entire argument being derived from his contention that the 'least' we can think being is as 'pure multiplicity,' the scientific thought of which is historically given by mathematics (in particular post-Cantorian set theory, which provides the minimal foundations for such a thought), which in turn leads directly to his famous equation of mathematics with ontology. Thus Badiou's 'subtractive ontology' is, properly speaking, a *minimal intrinsic ontology*.[34]

Yet, as Desanti points out, opting to think being minimally:

> in no way entails that ontology itself should be minimal. On the contrary, it is obliged to be maximal in at least this sense: so as to be sufficient for defining and circumscribing the regulated realm of that which presents itself as 'being' [*étant*] and as happening to being or coming into being [...]. What presents itself as minimal must then posit itself, in essence, as the basis for the maximal domain of all determinations locatable in the realm of beings.[35]

While the set-theoretic ontology of *Being and Event* certainly conceives of presentation and re-presentation (in the form of situations/sets and states/power sets), in restricting its terrain to 'the flat surface of indifferent multiplicity'[36] it equally limits its intellectual reach to what *presentation is*, and necessarily fails to grasp that which *is presented*. Or to put it another way, by reducing itself to questions of *essence*, of being in its 'integral transparency', subtractive ontology proves itself to be wholly incapable of accounting for how it is that multiples *appear*.

According to Desanti's criteria, such an ontology remains *too minimal*. Or again, while Badiou's intrinsic ontology is certainly minimal, it does not necessarily follow from this that it is *sufficient*: limited to its own immanent resources, set-theoretic ontology is unable to posit itself as the basis for the maximal domain of all determinations locatable in the realm of beings.

Without going into detail or pre-empting Badiou's own response – the delicate unfolding of which constitutes the bulk of this book – we will conclude simply by noting, once again, that Badiou seeks to answer this criticism not by turning 'inward' but rather by turning 'outward', that is, by supplementing his established set-theoretic ontology with a logical (category theoretic) phenomenology. Thus, with *Mathematics of the Transcendental*, Badiou moves from the intrinsic to the extrinsic, or from being *qua* being as the theory of the pure multiple, to a complementary theory of *being-there* (which is coextensive with being itself) as the topological localization of a being, or of its appearing in a world.

To this effect, Badiou's wager in *Mathematics of the Transcendental* is that he can move from the intrinsic to the extrinsic *without disturbing his set-theoretic ontology*. Or again, that a minimal intrinsic ontology can be supplemented by a *maximal extrinsic onto-logy*.

Such is the (categorial) imperative guiding this book, namely, that it is only by supplementing the minimal with the maximal, the intrinsic with the extrinsic, that a complete formal science of being-*there* might be established.

We leave it to the reader to judge his success for themself.

* * *

A work like this cannot be reducible to the one or even the two.

In 2010 the Melbourne Badiou Reading Group (est. 2005) turned its attention, finally, to *Logics of Worlds*. After several months we decided to suspend this folly so that we might first make the effort to 'know' category theory. We engaged a friendly mathematician, Callum Sleigh, and proceeded to read Robert Goldblatt's canonical *Topoi: The Categorial Analysis of Logic* line by line, a method that bemused our mathematician friend.

It was a valiant but vain effort. However, we had in our possession, via the good offices of Oliver Feltham, a facsimile copy of *Topos, ou Logiques de l'onto-logique: Une introduction pour philosophes, tome 1*. Being explicitly written 'for philosophers', we decided – with all the pretence this involves – that this would prove a more suitable initiation into the intricacies of category theory, and so began to translate pages for the reading group.

Let us be clear: none of us became instant experts in the theory. However, the clarity of the exposition, the patience taken to demonstrate and contextualize, and the demands made on us in the teaching, provided us with an anchor point in this 'extrinsic ontology', allowing us to hold this point and, from there, to proceed to hold others.

In this sense our experience of the book somewhat conforms to Rancière's depiction of Jacotot's pedagogical method, as, page by page, diagram by diagram, we proceeded to teach ourselves what we did not know. The same could even be said to apply to Badiou himself, for it is he, in the first instance, who is the philosopher being 'introduced' to category theory. Indeed, what we see in *Topos* is ultimately the formation of a philosophy, whose procedure of verification is *Logics of Worlds*.

The realization that an education in category theory was possible for us meant, of course, that it was possible for anyone, and we decided it was necessary to complete the task of translation. Badiou was immediately receptive and just as immediately sent us the second volume of this work, *L'Être-là: Mathématique du transcendental*. Having nearly finished Book I by this time, we were, it must be said, a little crushed. It quickly became evident however that Book II constituted, as Badiou had said to us, 'the logic of *Logics of Worlds*', and was as such the perfect complement not only to this decisive work, but also to the difficult investigations undertaken in *Topos*.

So we have here in this single volume, presented as such for the first time in any language, the necessary groundwork, together with the work that will condition the philosophy. It is without doubt an exemplary education in the work of philosophy or – it is the same thing – philosophy *as* education.

Of course, we cannot conclude without thanking the excitable members of the Melbourne Badiou Reading Group for their support in the project, and in particular Justin Clemens for his commitment, his intensity and his love of diagrams. We also thank Josh Comyn for his close reading of the text.

Bartlett would also like to thank Angela Cullip for her clarity and intolerance.

We reserve special thanks for John Cleary who, having spent two years in Paris studying mathematics at the Sorbonne and the École normale supérieure, returned to Australia and took up the task of correcting our French, our terminological failings, our mathematical knowledge and our hubris.

We would also like to thank Alain Badiou for his belief in us and his ongoing support for this and other projects, and we thank Isabelle Vodoz for her support too.

Finally, thanks must go to Haydie Gooder at Gooder Editing for another instructive index and to Rachel Eisenhauer and Liza Thompson for their patience in waiting for this tricky manuscript to arrive.

PART ONE

TOPOS, OR LOGICS OF ONTO-LOGY: AN INTRODUCTION FOR PHILOSOPHERS

1 GENERAL AIM

Eilenberg and Mac Lane introduced Category theory in the 1940s to serve the needs of algebraic geometry, just as set theory had been invented by Cantor 70 years earlier to address a number of problems of real analysis.[37] In both cases however the theory proved to hold such universal power that it developed into an entire mathematical ontology. This movement establishes the compact form of these theories as the means by which mathematics conditions philosophy.

In the dominant epistemological language, it is said that these theories are today rivals with respect to the (so-called) problem of the 'foundation of mathematics'. It would be far more appropriate to say that they define entirely different directions of thought regarding the prescriptions for the foundations of philosophy. So much so that the consideration of what is at stake in this opposition is an immediate and decisive problem for anyone who engages in the construction of the place of thought that is a philosophy.

The ontology prescribed by set theory determines being *qua* being as pure multiplicity 'without one', and the language of this ontology is sutured to the presentative position in general through the empty set as nomination. The ontology prescribed by category theory determines being as act (or relation, or movement). Its basic concept is the arrow (or morphism, or function) which 'goes' from one object to another object. We should not be misled by this objectual vocabulary: an 'object', in category theory, is in the first place a simple point (or a simple letter) without any determinate interior, while a set is precisely nothing other than the count-for-one of what belongs to it, namely, its 'interior'. It is for this reason that Desanti is right to say that set-theoretic ontology is 'intrinsic', while categorial ontology is 'extrinsic'. By extrinsic we mean that in a categorial universe an object is determined exclusively *by the relations, or movements, of which this object is the source or the target*. Even the categories of immanence (which are the 'elements' or the 'sub-objects' of an object) are defined by arrows which go (or operate) toward (or on) the object in question.

In set theory the identification of a presented multiple (therefore of *a* being) is clear: a set is defined by its elements and two sets that have the same elements are identical (are the 'same' set). This is prescribed by the axiom of extensionality. We see that this axiom rests on an *intrinsic* (immanent) determination of the pure multiple. In category theory this is generally not the case. Because 'two' objects, while counting-for-two 'in itself' (in a set-theoretical sense), can obviously be the source and target of the same kinds of actions (or relations) in the category in which they appear. In which case their *extrinsic* identity is the same. It is however this identity, and not the pure difference of position (which is in fact a literal difference: one of these objects is called a, the other b), that is essential for the categorician. In this case we will say that the two objects are isomorphic, and will regard them as effectively 'the same'. There is therefore in category theory an essential ambiguity to identity. Caught up in a network of relations and similar external actions, 'two' objects are indiscernible, except as pure empty letters. This is inevitable since identity is not an immanent mark, but rather the effect of actions, or arrows, which operate in a categorial universe.

To the extent that 'nature' (in the Aristotelian sense) is identified with the movement that actualizes the resources of a substance, we will say that category theory attempts to align ontology with a 'natural' conception of being. It is not for nothing that 'natural transformation' is one of the key concepts of category theory. Set-theoretic ontology on the other hand is a matter of un-natural intelligibility, since it builds the universe out of the void alone, and subordinates movement (the functions) to a *fixed* position of presentation (the being of a function is in fact its graph, that is, a pure multiple whose elements are pairs of the type (a, b), or $b = f(a)$). We will also say that the opposition of set theory to category theory is a contemporary incarnation of the opposition between Plato and Aristotle.

Just as Aristotle held that being is said 'in several senses', likewise a concept in category theory is, in its scope and in its components, entirely dependent on the categorial universe in which it is defined. This is inevitable, since a concept (or an object) is identified only through its movements, through the operations performed, movements and operations that involve the entire universe (all the other 'objects', but above all the 'forces' or arrows which criss-cross this universe). In set theory on the other hand there exists an 'absolute' universe of reference, namely, the cumulative hierarchy of sets, which is sutured to being *qua* being through the name of the void, and which opens up successive levels that can be traversed by thought on the basis of fundamental operations: by passing from a set to the set of its parts, or to the union set. Category theory is entirely relativist, it shows a plurality of possible universes (called Topoi). Set theory contrarily presents ontology in the unfolding of a unified intelligible world.

The procedure that founds set theory is the effect of a *decision regarding the intelligible*. This decision takes the form of axioms, which specify the crucial Ideas organizing the thinking of being *qua* being. The categorial procedure is itself *descriptive*. It proceeds by successive definitions, which enable the specification of possible universes. This opposition between axiomatic thought and definitional thought reiterates the opposition between Plato and Aristotle.

The 'natural' logic of set theory is classical, in that it admits the principle of excluded middle and consequently reasoning by the absurd. Ontologically, this refers to the fact that, since there is an absolute concept of being, *that which is not non-being must be*. The absoluteness of the determination of being is equally the absolute delimitation of the gap between being and nothingness. In category theory the relativist plurality of universes and the 'extrinsic' character of immanence implies that the negation of non-being is not, in general, equivalent to the affirmation of being. This means that intuitionism is the natural logic of category theory, and that reasoning by the absurd is inadmissible in many Topoi: to show in any given universe that a particular object 'exists', we must somehow *show* it (that is, 'construct' it, as always, from the arrows of which it is the source and the target); it will not suffice to establish that its non-existence is contradictory. There is therefore in this kind of thought no indirect (or oblique) proof of existence or truth. We are only bound to believe what we see (or construct). Hence an empiricist bent, broadly speaking, once again opposes itself to the exclusively intelligible constraints of set theory, which, separated by the void from any presupposition of presence, makes extensive use of indirect proof, which shows nothing but teaches that what exists for thought is altogether beyond what experience is able to present.

Finally, the 'normal' position of category theory with regard to the infinite is one of reticence. Certainly, categorial universes are 'large', even larger than what is tolerated by set theory. Thus we admit the 'existence' of a category of sets while, as we know, it is contradictory to posit the existence of a set of all sets. But *from the interior* of a category, subordinated here to an intuitionistic logic, the handling of the infinite is limited. We can clearly see here the free use that set theory makes of the axiom of choice (which is in fact an existential axiom of infinity: it posits the existence of a set which 'chooses' *one* element from each set of a given collection of infinite sets). In category theory, this free use is limited by the fact that a good many Topoi do not admit the axiom of choice (we will come back to this essential point: for if a Topos admits the axiom of choice, its logic *must* be classical and not intuitionistic).

Set theory and category theory therefore offer distinct paths for all the decisive questions regarding the thinking of being (acts of thought, forms of immanence, identity and difference, schools of logic, admissible rationally, relation of experience to existence, infinity, unity or plurality of universes, etc.). This is to say that they *lay out different conditions for philosophy*. We have here a debate crucial for the construction of the space of philosophy that invariably takes the form of a matrix of ontological choices.

2 PRELIMINARY DEFINITIONS

A category is defined by its *objects* (written a, b, c, ...) and its *arrows* (written f, g, h, ...). Each arrow 'goes' (or operates, or acts) from one object (its source, or its domain) toward another object (its target, or its co-domain). An arrow is thus a *directed* action:

$$a \bullet \xrightarrow{f} \bullet\, b$$

The use of schemas (or diagrams) is essential, because they always *show the shape of a definition spatially*. The diagram corresponds to the natural and intuitive conviction of categorial ontology, and demonstrates 'in a single stroke' ('uno intuitu', as Descartes would have said) the presentative consistency of its definition. We note that by contrast, most schemas of sets, beginning with the famous Venn diagram which turns a set into a two-dimensional 'potato', are quite misleading (it is immediately impossible to visualize a transitive set, namely, a set whose elements are also its parts).

Note: Be careful to observe that between two objects there can be many different arrows. For example:

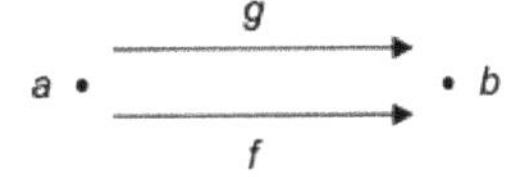

What does it mean to say that the arrows f and g are different? In general, it will mean that they 'combine' with other arrows in different ways. Here again, the principle of same and other must remain extrinsic.

If an arrow f has as its target the object b, and an arrow g has as its source the same object b, this means that one arrow 'acts on' b, while another action 'leaves' from b. It is eminently reasonable to think, then, that an action exists which 'links' the first to the second, an action which is composed first of the action f, then the action g, with the object b as the 'junction' of these two actions.

In this case, we propose the existence of a 'composite' of f and g, written $g \circ f$:

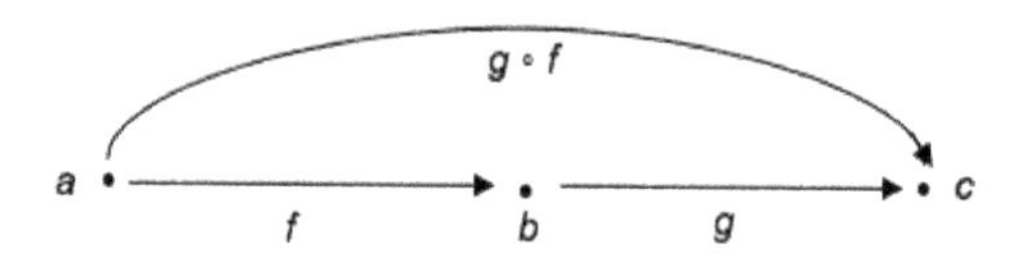

Note: Writing $g \circ f$ indicates that f acts before g; it must therefore be read from right to left. This is because, when dealing with functions in the classical sense, we very naturally write $g[f(x)]$ to indicate that g (for argument's sake) is obtained from x by f. Though it does take some getting used to.

The minimum requirement for calculating the composition of arrows, the 'natural' requirement, is as follows: if you perform the action f, followed by the action g, you get the action $g \circ f$. If you now perform the action h, applied to the composite action $g \circ f$, you have a composite action $h \circ (g \circ f)$. But you can also perform the action f, *then the composite action* $h \circ g$, giving you the total action $(h \circ g) \circ f$. Are these results identical? Are they, in the end, the same action? Since it is *only a scansion*, a punctuation, a chance to catch your breath after f, rather than after the concatenation $g \circ f$, we consider the final result of these three linked actions to indeed be the same, namely:

$$h \circ (g \circ f) = (h \circ g) \circ f$$

Technically, this is said: the composition of arrows in a category is *associative*.

The prevalence of arrows over the 'identity' of objects takes the following form: We consider, in category theory, that each object is the support for *a null action that goes from itself to itself*. How do we define a null action? Once again, extrinsically. Suppose that an object a is the target of an arrow f, and the source of an arrow g. The null action from a to a is such that if it is made after f, or before g, it is as if you *only* performed f (or g). To put it another way, if we call this null arrow **Id**(a) (for 'identity of a'), we will have:

$$\mathbf{Id}(a) \circ f = f$$
$$g \circ \mathbf{Id}(a) = g$$

Further, when it enters a composition, the null arrow 'counts as nothing'. It is a neutral element of the composition.

Schematically:

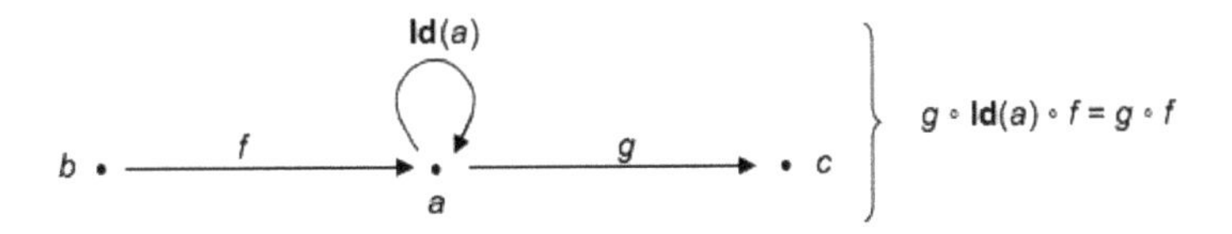

Ultimately, the identity arrow *is* the object a, considered as the 'halting point' of the action. The somewhat Bergsonian (or Deleuzian) idea underlying this is that an identity is only ever a cessation of movement, or the absence of movement.

> *Note*: We will be careful to observe that there may be any number of arrows between a and a which *are not identity arrows*. What characterizes identity arrows, as always in categorial thinking, is not its position (it goes from a to a), but its behaviour within the compositions of actions or arrows (in this case its 'nullity'), and therefore the *relationship* that it maintains with anything other than itself. You may very well have an arrow f whose source is a and whose target is a such that we have:

$$b \xrightarrow{g} a \xrightarrow{f} a \qquad f \circ g \neq g$$

In this case f is different to **Id**(a), and you will notice that 'different' means: giving another 'result' *when we compose it with* g. This maintains the concept of difference within a protocol of extrinsic evaluation (to know that f is different to **Id**(a), we need a third term, outside of f and **Id**(a), which is the arrow g).

A category consists of objects and arrows, provided that, given two arrows: there always exists the composite of these two arrows; this composition is associative; and for every object we have an identity arrow, which is neutral in any composition in which it operates.

As we can see, this definition is particularly 'lean' and startlingly general. It contrasts with the axioms given by set theory, which are from the outset extremely dense in ontological and conceptual content.

Note: There is of course no requirement that a category include arrows between all its objects. It may well be that no action links these objects, that between a and b there is no arrow whose source is a and whose target is b, or whose source is b and target is a.

There exists an object-less category, the empty category (it is also without an arrow, since all arrows must have a source and a target).

By contrast, even if a category has only one object, it must have at least one arrow, namely, the identity arrow of this object.

This gives us (after the empty category) a very 'small' category: the category with a single object, and a single arrow: the identity arrow of this object:

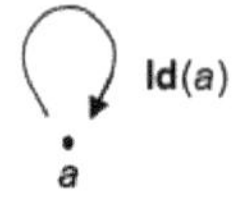

We call this category One.

3 THE SIZE OF A CATEGORY

The fundamental idea of categorial thought is that of defining possible universes. For a category to be a universe, it is essential that it be a sufficiently large 'space of actions'.

There is no way to determine 'in itself' the size of a category. This also contrasts sharply with set theory, because in set theory there is a general scale for the intrinsic quantity of multiples: the series of finite and infinite ordinals and cardinals. If we want a set that is a universe for a given statement (meaning: it makes sense to ask if this statement is true or false in this universe), we must in general consider *the cardinality of the universe*. In other words, where is it situated in the cumulative hierarchy of sets that unfold themselves, level by level, from the void? This is a straightforward question, since the levels of multiple-being are indexed onto the series of ordinals. This is not so clear in category theory. What matters here is knowing *which actions are possible in this universe*, and which *compositions* of actions in particular. The size of a category becomes a determination of the *possible*.

As you might expect from an extrinsic ontology, the suitable metaphors for thinking this type of approach are 'visual' or geometrical. A category will be all the more 'large' when we can 'see' from a point of the category the vast configurations located elsewhere in the category (rather like a room is large if you can shoot a scene taking place in the room without having to leave). We will say that a category is large enough if, from within the category, we can 'operate' with some perspective. To treat these metaphors rigorously, we must specify what a configuration (or a scene) can be *within* the category under consideration. This is very simple: we consider that we have a piece of the category when we have some objects, and (possibly) a certain number of arrows between these objects. This is called a *diagram*.

Note: A diagram doesn't necessarily include all the arrows which are in the category between two objects that are in the diagram. You can have the following diagram:

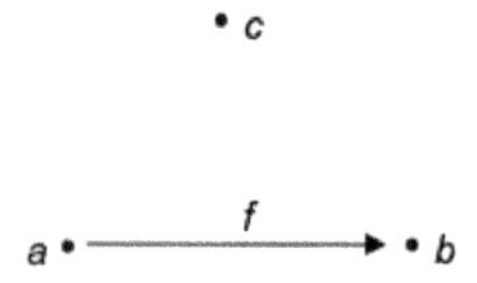

This is not to say that the arrow f is, in the above category, the only arrow existing between a and b, or that there is no arrow between c and a, or c and b. Simply, the categorial piece that is this diagram has taken the arrow f between a and b and no others (if there are any).

A diagram is not a category. It is any given piece, or an arbitrary piece, of a category. In particular, it need not follow that just because a is in the diagram, the identity arrow **Id**(a) – which necessarily exists in the category – is also in the diagram. Equally, if f and g are composable (that is, if the target of f is the source of g), the composite $g \circ f$ necessarily exists in the category, though it need not be in the diagram.

Finally, we maintain that for every category there exists the *empty diagram*, which presents no object (and therefore no arrow).

Arbitrary as it is, a diagram is in general very loosely structured. The number of compositions which are feasible within it can be quite small. To get a measure of the 'consistency' of a diagram, it is useful (in fact, absolutely crucial) to know what a 'very' structured diagram can be.

Up to this point we have only defined a kind of 'structuring' operation: the composition of arrows. Consider the following triangular diagram:

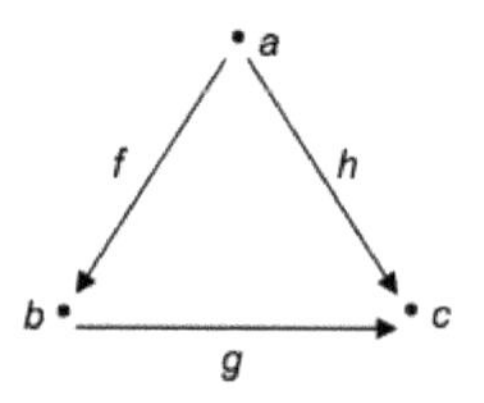

By the definition of a category, we know there exists *within the category* the arrow $g \circ f$. Is it in the diagram? Note that the composite $g \circ f$ is an arrow which goes from the object a to the object c. However, in the diagram there is indeed an arrow which goes from a to c, which is the arrow h. But:

> *Note*: We know that between two objects there may be entirely different arrows. Therefore, the arrow h, which goes from a to c, is not necessarily identical to the arrow $g \circ f$, which also goes from a to c. We could have:

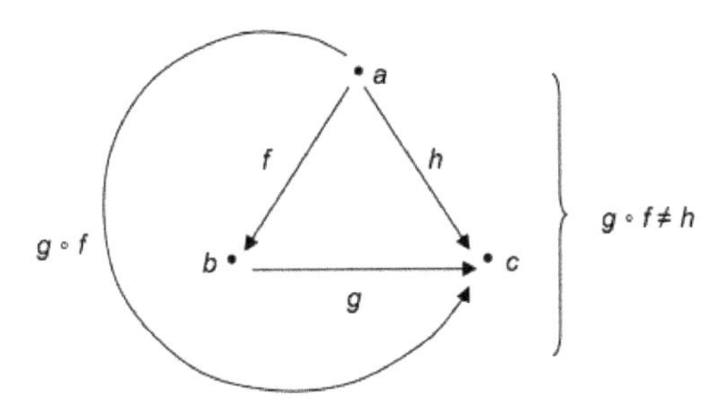

If the arrow h in the diagram is indeed the same as the composite $g \circ f$ then we say that *the triangular diagram commutes.*

The concept of the commutativity of a triangular diagram, though unimpressive in its appearance, is absolutely fundamental. Why? Because it designates, in terms immediately appropriate to the definition of a category, what a *structured* diagram is. A structured triangle is such that if you connect the action of both of its sides, you get *the same thing* as if you perform the 'unique' action marked by its base. A structured triangle is in fact a *double marking* of the same action, according to its decomposed form (the action h, which is the action $g \circ f$). This point is, for categorial thought, far-reaching. Ultimately, a categorial piece (a diagram) is consistent (or not completely arbitrary) if all the paths from one object to another contained therein are identical (are the same action). In other words, it is consistent if there are multiple marks for the same action, given either directly or by compositions.

Note that the fact that the triangle below commutes:

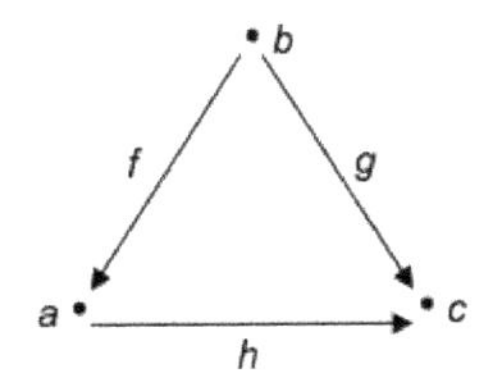

is the 'geometrical' expression of the simple equation $g \circ f = h$.

This is entirely in the spirit of categorial thought, which, true to Descartes, places mathematical exposition in a constant ambiguity between the algebraic (equations) and the geometric (diagrams). This ambiguity will only be fully elucidated once we establish that, ultimately, category theory is essentially an *intuitionistic* logic (as indeed is Cartesian logic). Which is to say that it organizes thought around actual constructions, which can in effect either be explicit equations or 'spatial' configurations.

We will say that a diagram is commutative if all its triangles are commutative.

Since the problem of the size of a category is that of the 'perspective' that we can have with regard to a diagram, it is simply a question of defining the metaphor of 'seeing', or of perspective, consistently, thus with consistent diagrams.

The requisites will be the following:

a Given the *objects* of the diagram, there exists an object of the category from which all the objects are 'visible'. Thus (since seeing, as with every action, must be thought with arrows) if a_1, a_2, ... are the objects of the diagram, and if c is the object which gives us 'perspective', there exists arrows which go from c to a_1 (let's call this g_1), to a_2 (call this g_2), etc. Thus we have the following diagram for c:

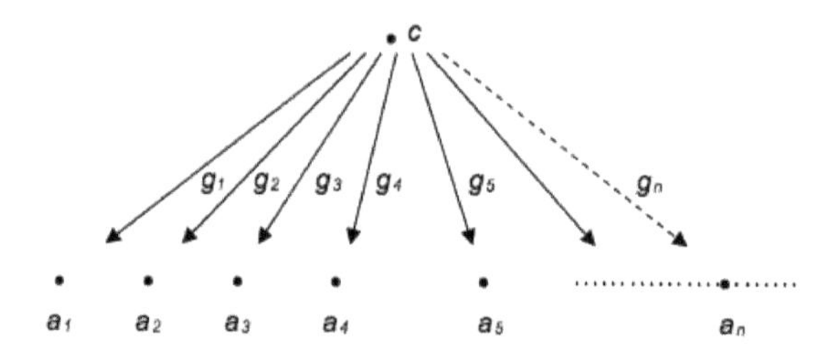

Note: c is an object *of the category*. It is not, in general, an object *of the diagram*, like a_1 or a_2.

b Given an arrow of the diagram, for example f between a_1 and a_2, the triangle whose peaks are a_1, c and a_2, and whose sides are the arrows g_1, g_2 and f, *commutes*:

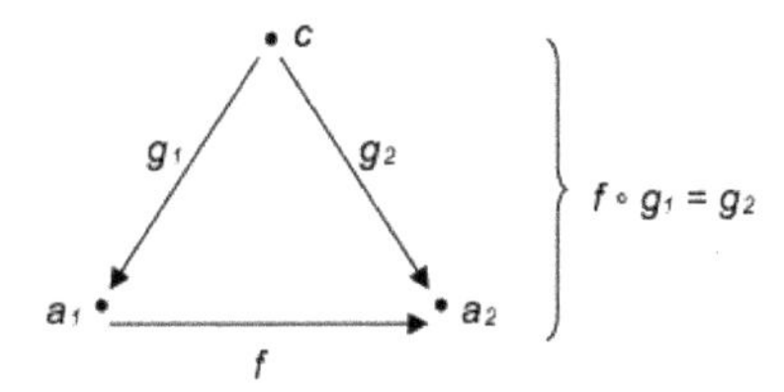

This second condition gives what might be called the consistency of 'vision'. Not only are all the objects of the diagram visible from a point of the category (the point *c*), but the arrows of the diagram are 'captured' by this view, in the sense that every arrow of the diagram *enters into the composition of an arrow coming from c*. In effect, as shown in the (commutative) 'visibility' diagram, we have: $g_2 = f \circ g_1$. So the arrow *f* of the diagram is 'analysed' by the arrows g_1 and g_2, which are the 'lines of sight' coming from *c*. We might say that, in this case, the system of point *c* and arrows $g_1, g_2, \ldots$ is an *analytic view* of the diagram.

This system is called a *cone*. The diagrams adequately show why it is given this name. A diagram in a category *admits a cone* if there exists an object *c* in the category together with the arrows $g_1, g_2, \ldots$ from this object toward all the objects of the diagram, such that the corresponding triangles commute.

A category will be 'large' if many diagrams, and therefore many pieces of the category, admit cones (are analytic views of the interior of the category).

4 LIMIT AND UNIVERSALITY

A diagram can of course admit several different cones (we 'see' the diagram analytically from several objects, or points, of the category). Can we isolate, amongst these points, *one of them* as a point of 'universal' vision?

The concept of universality at play here is typically 'categorial', in that it is geometrical. With respect to a diagram, it is a universal *position* that we are looking for. That is, a point where we see the diagram *as closely as possible.*

What might 'closely as possible' mean? The idea is that the 'universal' point *is itself visible from the other points from which we see the diagram.* It is like the outpost of all the points of the category from which we see the diagram. Technically, this means (rules of the analytic view):

a that if c is the universal cone-point from which we see the diagram D, and if c' is yet another cone-point of the diagram, there exists an arrow h from c' to c (c is 'visible' from c').

b that if g_1 is the visibility arrow which goes from c toward an object of the diagram a_1, and if g_1' is the arrow which goes from c' toward the same object a_1, the triangle whose peaks are c, c' and a_1 and whose side arrows are h, g_1 and g_1' *commutes*:

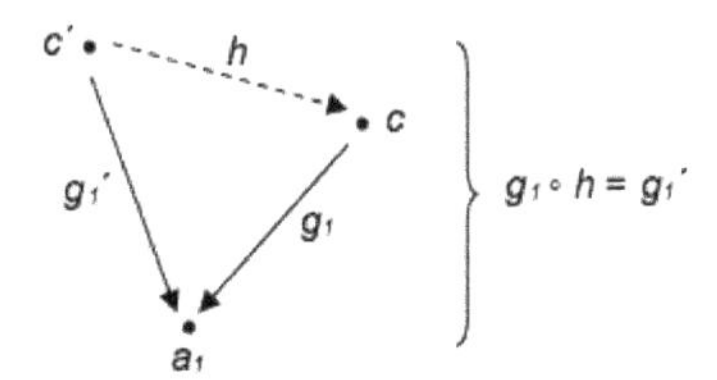

This second condition establishes that the arrow g_1' is 'captured' by the universal point c, since the (commutative) diagram shows that we have: $g_1' = g_1 \circ h$, and that g_1 is a source arrow from c, while h is a target arrow of c.

To complete the universality of cone c however, obviously a condition of *uniqueness* is needed. Being in a universal position with regard to a property (here, the analytic vision of a diagram, although in category theory this is the matrix of *every* property) engages a positional effect, whose principle we outlined above, *and* a principle of uniqueness. All universality is a composite of subsumption and of the One. However, the One in category theory cannot be the One of *the object*. It must be the One of *an action*. So we say that the arrow h, which subordinates the cone c' to the universal cone c, is *unique*. This is precisely to say that *there is only a single arrow* from c' to c which makes the triangular diagram c'-c-a_1 commute (and this of course equally goes for all the other objects of the diagram).

Given a diagram D, we say that it admits a limit if it admits a cone c such that, for every other cone-point c′ of the same diagram, there exists a unique arrow h from c′ to c which makes all the triangles of the kind c′-c-a commute, where a is an object of the diagram, and where the arrows from c to a and from c′ to a are visibility arrows prescribed by the existence of cones.

We say that the limit cone of the diagram D has *the universal property* with regard to this diagram.

This concept is absolutely central. It determines the universal not by extensionality (the 'for all' which refers to a set), but by the combination of a geometrical or positional property (the analytic view of the cone), and a property of uniqueness directly attached to the actions that the universal supports *from those points which share its positional property.*

This being so, is there no 'objective' uniqueness to the universal (or of the limit)? There is one, *in the categorial sense.* We demonstrate that for a given diagram D, *two objects possessing the universal property are isomorphic.* We will see the precise definition of an isomorphism later on, but it will suffice us here to say that two limits (this being the same thing as the universal property) are not extrinsically discernable, that is, discernable by the 'active' properties (or in terms of arrows) which determine them, even if they are so 'literally'.

5 SOME FUNDAMENTAL CONCEPTS

The principal concepts of category theory will henceforth be defined *as limits*. This is hardly surprising, given that 'limit' means 'universal position'. The geometrical substrate is obvious: we start from a diagram, thus from a configuration in the category, and ask: is there an object that possesses the universal property for the diagram? Such an object, and the system of actions (or arrows) which identify it, will receive a conceptual name. A category will be gradually *described* by the existence (or not) of objects that fall under such concepts. In particular, the greater the number of limits for every kind of diagram, the 'larger' it will be. The extension acquires a descriptive or intentional sense (to support a wide range of conceptual names, or to provide, from within the category, the *cases* for these names), and not, as in set theory, a directly extensional sense (the cardinality of the universe). Categorial thought thus tends, at the heart of ontological presentation, to reduce quantity to quality.

a) Limit of the empty diagram: terminal object

Let's start with the 'object-less' diagram. What is a cone for this diagram? Since there is nothing to 'see', it is obvious that *any object* in the category concerned has the capacity for such a 'view' (which is the capacity to do nothing). The empty diagram, for any category, admits any object of this category as its cone.

What then is a limit for the empty diagram? It is an object which is 'visible' from every object, since they are all cones. Moreover, the visibility arrow must be unique (condition of universality).

So: in a given category, the empty diagram admits a limit if there exists in the category an object, denoted 1, which is such that *there exists, for every object of the category, one and only one arrow which goes from this object toward 1*. Let's agree from hereon in to mark with a dotted line an arrow that obeys a condition of uniqueness. Thus for every object a of the category the diagram:

The object 1 is called a *terminal object*. If a category admits a limit for the empty diagram, we say that it is a category 'with terminal object'.

b) Limit of the diagram with two objects without arrows: product

Given the diagram made up of two objects a and b, with no arrows:

A cone for this diagram is a point c' from which two arrows proceed toward a and toward b (where a and b are 'visible' from c').

A limit for this diagram is a point c from which two arrows also proceed equally toward a and toward b, but which is also 'visible' in a unique way from any other point possessing this property, for example from c'. It is also necessary that the triangles c'-c-a and c'-c-b commute. The diagrammatic situation is therefore the following:

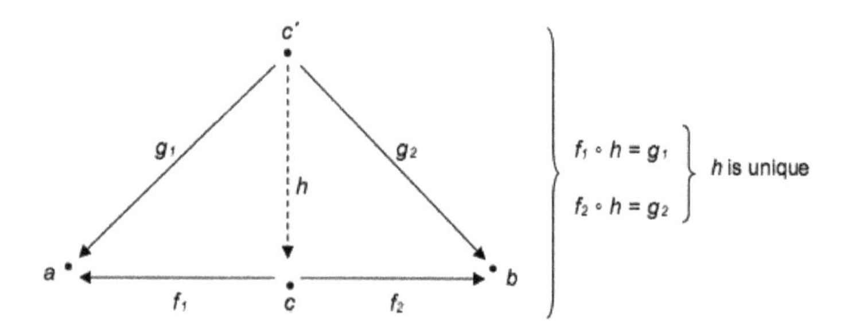

We call the limit of this diagram made up of the objects a and b the *product* of a and b. We write $a \times b$ for the limit-point c. The arrows which go from $a \times b$ to both a and b are called *projections*. The unique arrow which goes from c' to $a \times b$ is called the *product map* of g_1 and g_2, which we write as $\langle g_1, g_2 \rangle$. To redo the diagram:

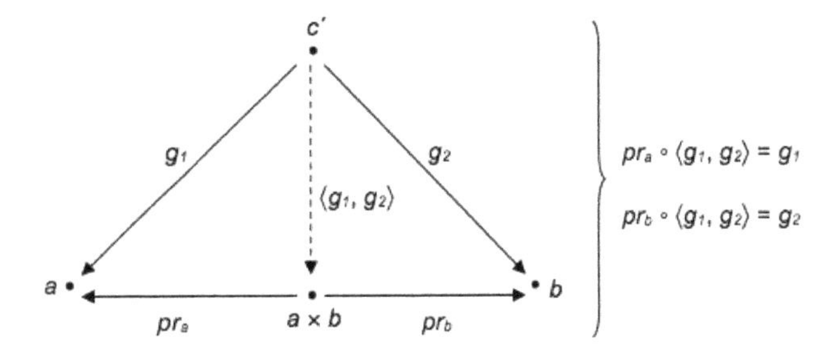

Pay close attention: there exists, for a and b, an object $a \times b$ such that there is an arrow from this object toward a and an arrow toward b such that for every object c' which is also the source of two arrows toward a and b there exists a *unique* arrow from c' to $a \times b$ which makes the diagram commute. If this is the case for every pair of objects (a, b) in a category, we say that the category *admits products.*

c) Limit of the diagram with two parallel arrows: equalizer

Take the following diagram:

$$a \underset{f}{\overset{g}{\rightrightarrows}} b$$

In the geometrized vocabulary of category theory, f and g are said to be parallel arrows. A cone for this diagram is an object c' which sends the arrows toward a and b, so that the resulting diagram commutes:

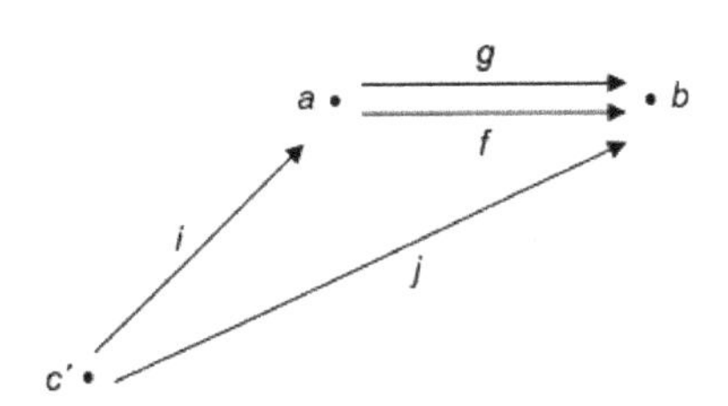

Note: Since the diagram commutes, you have (for example): $j = g \circ i$. As the arrow g is given initially (in the diagram in which we sought the limit), the arrow j is *entirely determined* by the element i (it is the composite of i and g). We can therefore, in our search, simplify the diagram, and only mark:

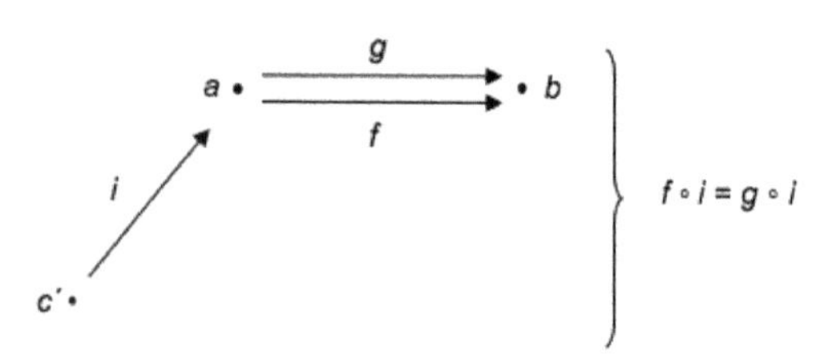

We must get used to these 'flattenings', or spatial modifications, inasmuch as the *only* thing that counts is the legibility of the compositions of arrows.

Finally, the cone is obtained when have a point c' with an arrow i from c' to a such that $f \circ i = g \circ i$. This is why we say that the arrow i *equalizes* the arrows f and g. By making the action i occur 'before' the (generally speaking different) actions f and g, you obtain identical composite actions. The action i 'reduces' in advance the difference between f and g.

Now, a limit will be a universally positioned cone, thus a point c and an arrow h to a where $f \circ h = g \circ h$, such that a unique arrow k comes from c' (as for every other object in cone-position for the diagram with two parallel arrows) that makes the diagram commute:

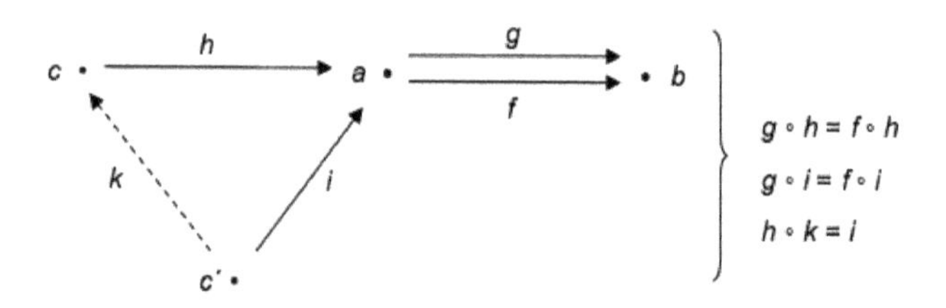

You will remark that the commutation amounts to saying that $h \circ k = i$, therefore that i is 'captured' (or factorized) by h (coming from the limit c) and k (the unique arrow which specifies the universal position of c).

Such a limit will be called an *equalizer* of f and g. If a category possesses a limit for every pair of parallel arrows, we say that the category *admits equalizers*.

d) Limit of the diagram with two arrows that have the same target: pullback

Take the following diagram:

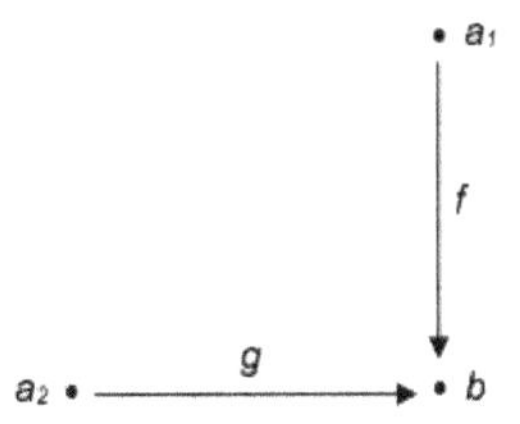

A cone for this diagram is immediately given by the diagram:

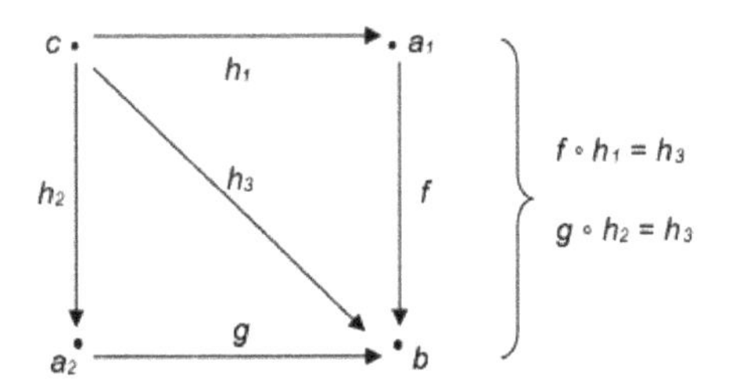

> *Note*: Just as we remarked in the case of the equalizer, the arrow h_3 is entirely determined by the composition $f \circ h_1$. Thus we can simplify the diagram and mark:

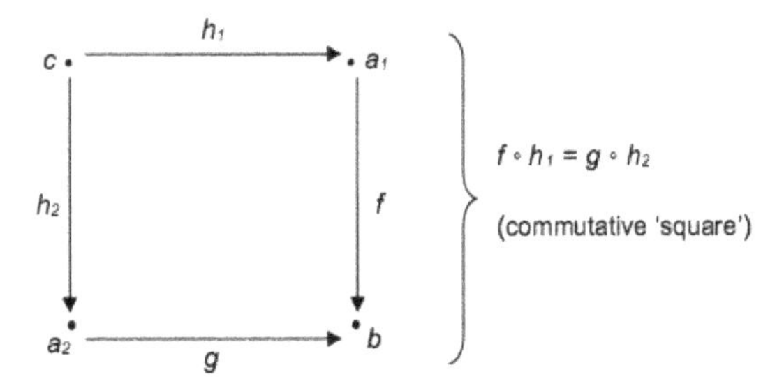

A limit is then a cone-point c which, for every cone c', gives the following diagrammatic situation:

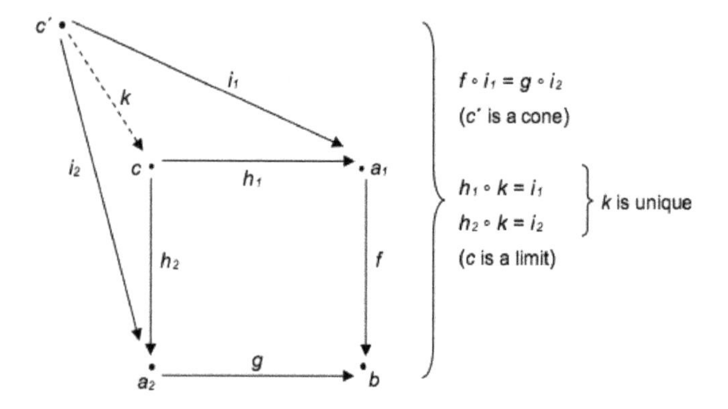

Such a limit is called a *pullback*. If every pair of arrows with the same target has a limit, we say that the category *admits pullbacks*.

e) Key concept: monomorphisms, or monic arrows

Consider the following diagram:

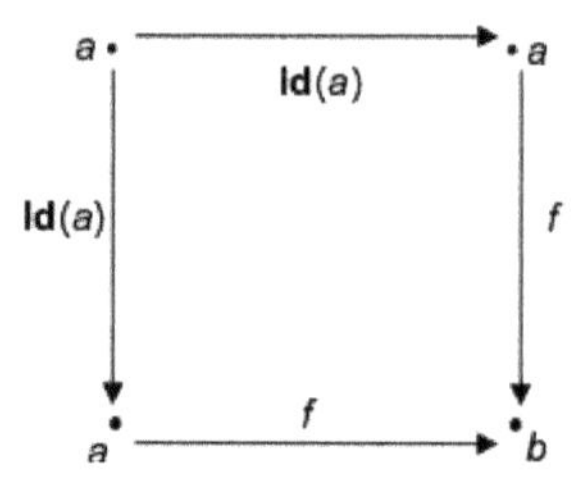

Note 1: Notice that there is a kind of geometrization of the impossible at work here, since the arrow *f* is marked 'at a distance from itself', or rather, the diagram inscribes the identity of *f* in two distinct places. We mustn't forget however that in category theory *geometry is the fixed layout of an action (or of a movement)*. The underlying thought is 'temporal': the action *f* takes place 'two times'. We will return presently to this categorial 'temporalization' of concepts, together with Kripke semantics (true statement 'at a certain point in time') and 'intentional' sets, where, throughout a certain 'time', *e*'s belonging to *E* may be a mere possibility.

Note 2: The diagram above always exists in a category (if, at any rate, it includes at least one arrow). Why? Because the identity arrow **Id**(a) necessarily exists.

Now, let's suppose that this diagram *is a pullback*. Or to put it another way, the object a and the arrow **Id**(a) are limited by the diagram:

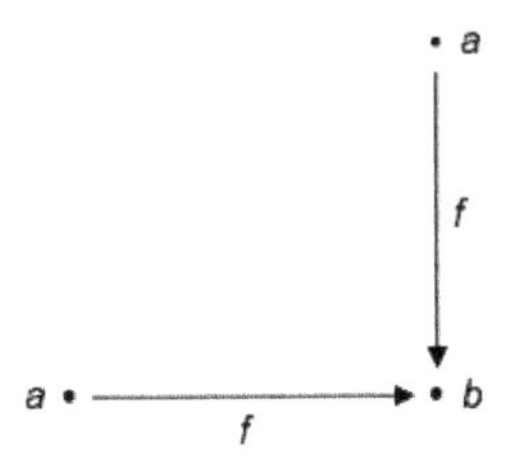

This means that for every object c which admits arrows g and h toward a and the appropriate triangular commutations, we get the following diagrammatic situation:

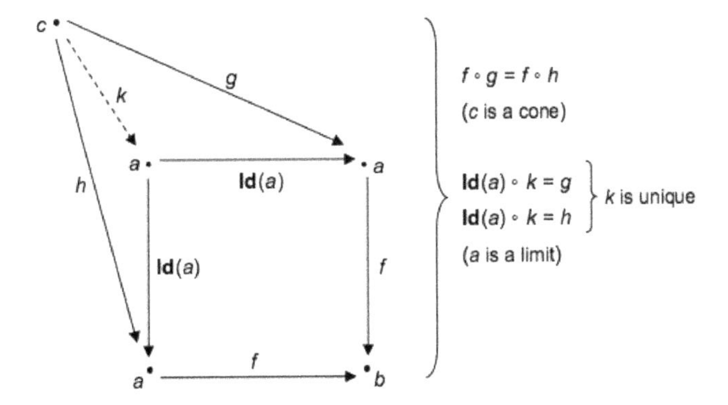

Given that the diagram commutes, we have **Id**(a) $\circ$ $k = h$. But by the definition of **Id**(a) (null action), this means that $k = h$. In the same way (the other triangle of the commutative triangle) we have **Id**(a) $\circ$ $k = g$, therefore $k = g$. Thus we finally have $h = g$.

Otherwise said: if we have $f \circ g = f \circ h$, we necessarily have $g = h$. We can also say that f is 'left-cancellable'. We can present the diagram in a flattened form:

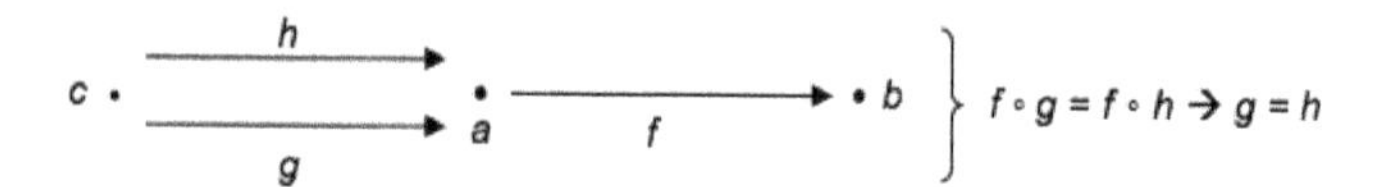

There is another sense in which we can take this: if h and g are *not identical*, then $f \circ g$ and $f \circ h$ can no longer be. This is somewhat contrary to the equalizer: if you perform the action f 'after' the actions g and h, you only get the same action *if the actions g and h were already the same*. The essence of the action f is that of *preserving the marked differences in the object* a *through the actions (from target a) g and h.*

An arrow f that has this property (which we have defined, as we should, in terms of a limit, thus of a universal property) is called a *monomorphism* or a monic arrow.

A monomorphism is an action that carries difference as such: it cannot install identity there where the preceding actions had instated difference. This operator is decisive, insofar as it regulates the question of identity and difference 'in movement'.

6 DUALITY

Take a given category **C**. Now consider the category obtained by inverting **C**'s arrows. This means that if in **C** you have an arrow f from a to b, you consider the new category to have an arrow from b to a. It's a little like if you place the category in front of a mirror, which is why we denote the new category $\mathbf{C}^{op}$, where 'op' stands for 'optical' or 'opposite'. $\mathbf{C}^{op}$ has the same objects as **C** and just as many arrows, only $\mathbf{C}^{op}$ arrows act 'in the opposite direction'.

Of course we must demonstrate that $\mathbf{C}^{op}$ is indeed a category (existence and associativity of arrow compositions; existence of identity arrows). This is a good exercise.

The category $\mathbf{C}^{op}$ is said to be the *dual* category of **C**. The workings of duality – of the inversion, of the addition – are essential here. Category theory is especially suited to the examination of ontologically 'dual' situations, which is to say, of reversible correspondences, of positional ambiguities, of inverted identities, or of symmetrical and mirroring effects. In this sense too the spontaneous philosophy of category theory is Deleuzian, coming as close as possible to symbolizing the imaginary.

When we move from **C** to $\mathbf{C}^{op}$, all the concepts are 'reversed'. Let's take a simple example: what does a cone of **C** become in $\mathbf{C}^{op}$? In **C**, there is a point c from which we 'see' all the objects of a diagram. Thus there are arrows from c toward all the objects of the diagram. This means that in $\mathbf{C}^{op}$ *there will be arrows from all the diagram's objects toward* c. It is the point c that 'is seen' from all the objects of the diagram. So while the commutations will be preserved, they will be turned 'in the other direction'. Such a situation is called a *co-cone*. For every cone of **C** there corresponds a co-cone of $\mathbf{C}^{op}$.

This is exemplified in the diagram made up of a single arrow:

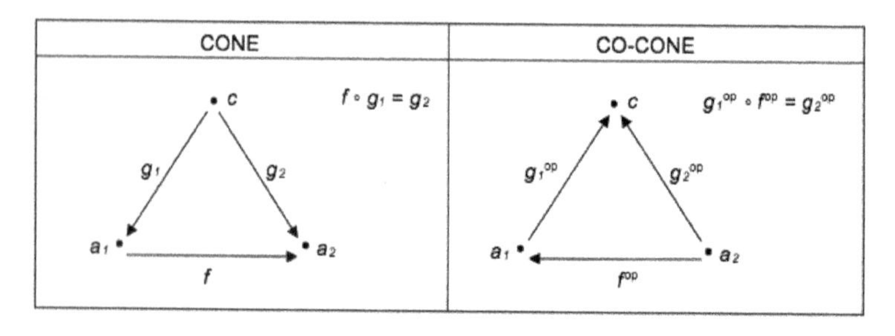

What then becomes of the limit? Since the limit c in **C** is 'seen' from all the cone-points c', in $\mathbf{C}^{op}$ it is all the cone-points that are seen from the limit. In this sense the limit in **C**, being an object that analytically 'sees' the diagram as closely as possible, becomes in $\mathbf{C}^{op}$ an object that is seen by the diagram from *as far away as* possible. Like this:

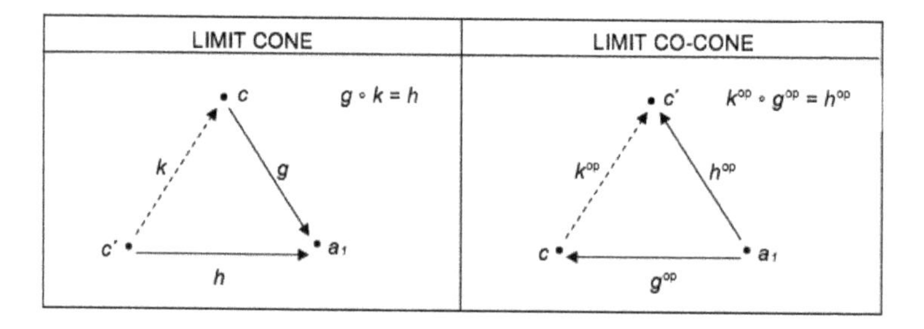

Such a situation will be named a co-limit.

Let's now take a terminal object in **C**. This is the limit of the empty diagram, an object 1 such that there is a unique arrow from every object of the category that goes toward it. In $\mathbf{C}^{op}$ we will have the existence of a co-limit for the empty diagram, which will be an object such that *there is a unique arrow going from it toward every object of the category*. Such an object (if it exists in a category) is called an *initial* object, and is denoted 0. We get the diagram:

Thus all our fundamental concepts will have a dual concept. I'll indicate this further through contrasting (and obviously commutative) diagrams.

a) Dual of the product: co-product (or sum)

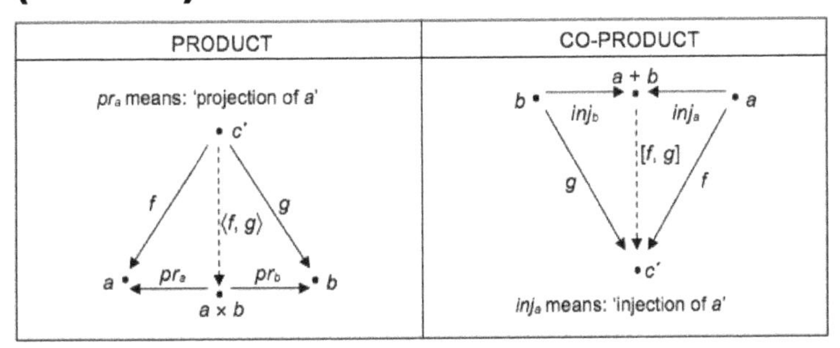

b) Dual of the equalizer: co-equalizer

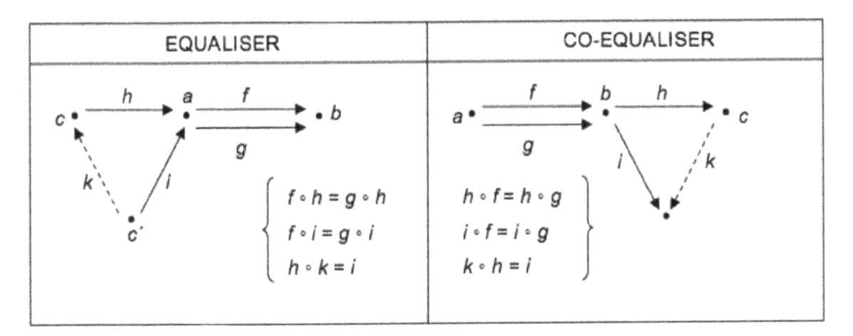

c) Dual of the pullback: pushout

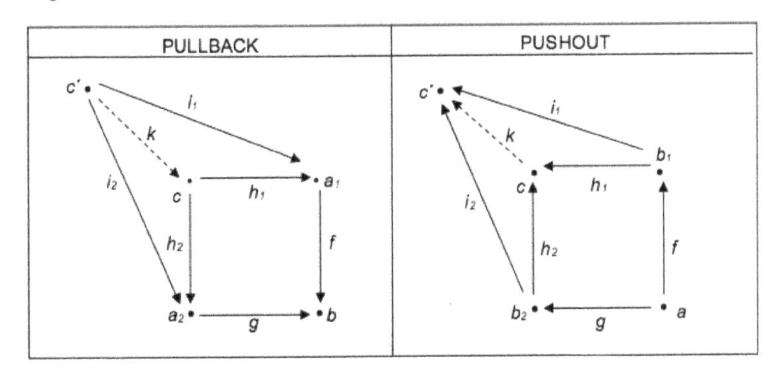

Special mention must be made here of the dual of a monomorphism.

An arrow is a monomorphism if, acting *after* two parallel arrows, it preserves difference. So if $f \circ g = f \circ h$ this implies that $g = h$. By duality, we will have this: an arrow f which, acting *before* two parallel arrows g and h, is such that if $g \circ f = h \circ f$ then $g = h$. See the two diagrams:

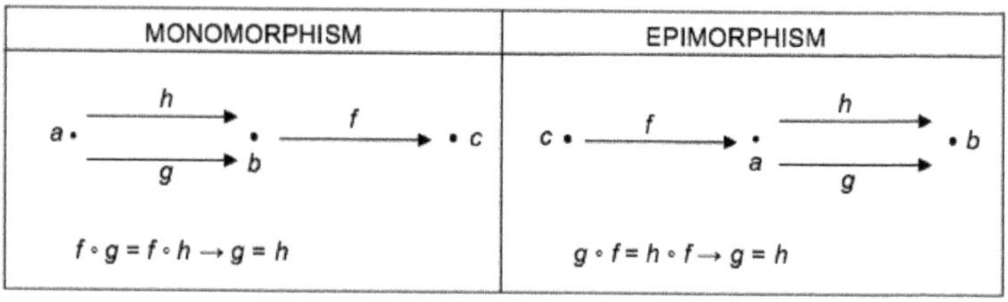

MONOMORPHISM	EPIMORPHISM
$a \underset{g}{\overset{h}{\rightrightarrows}} b \xrightarrow{f} c$	$c \xrightarrow{f} a \underset{g}{\overset{h}{\rightrightarrows}} b$
$f \circ g = f \circ h \rightarrow g = h$	$g \circ f = h \circ f \rightarrow g = h$

Such an arrow is called an *epimorphism*, or an epic arrow.

An epimorphism is an action such that, if it affects an object a, two actions which start at a can, when composed with the first action, give two identical effects only if they were themselves identical. Just as a monomorphism cannot introduce identity as it preserves differences, likewise an epimorphism cannot introduce difference, as preserves identities. Which is to say that an epimorphism does not inscribe any difference in a such that two different actions that begin at a would ultimately produce an identity. All difference must come from the dual actions following an epimorphism. Intuitively, this means that an epimorphism affects the object a 'outright' (or entirely), so that when you go from a to the actions g and h, you will ultimately end up with difference *provided that* it is induced by the difference of g and h. And if g and h are indeed different, you cannot, by the anticipatory action of the epimorphism, secure identity.

Thus monomorphism and epimorphism address the question of identity and difference 'in movement' *inasmuch as they are induced by actions*, according to the before (epimorphism) or the after (monomorphism).

The algebraic transcription of this geometric and temporal intuition is that, in the composition of arrows, a monomorphism is left-cancellable ($f \circ g = f \circ h \rightarrow g = h$) while an epimorphism is right-cancellable ($g \circ f = h \circ f \rightarrow g = h$).

What addresses the question of the categorial identity of *objects* is isomorphism.

7 ISOMORPHISM

How, within a logic founded on actions (or ordered movement), is the 'structural' similitude of two objects determined? The idea is quite simple. Take two objects a and b. Suppose that there is an arrow f from a to b. This means that an action originating from a 'affects' b, leaving a singular mark. Now suppose there is an arrow g from b to a (thus in the opposite direction) *which, when linked to the action f, produces a null action.* And the same applies if you act in the other direction: you go from b by way of g, then you return to b from a by way of f, and it's as if you haven't done anything. Obviously, under these conditions, the objects a and b are 'actively' identical, in that the singular marking of b by the arrow f from a allows itself to be cancelled out by an action leaving from this mark and returning to a. The two arrows (f, g) establish between the objects a and b a kind of 'inactive' correspondence, which, as always in category theory, is the sign of identity (identity as the halting of movement, or as movement without effect).

The primary paradigm for every null action is the identity arrow attached to every object. The situation we seek to describe will obviously engage this paradigm. If we have f going from a, and then we have g going from b to a, we get the composite $g \circ f$ *which is an arrow from a to a.* To say this is a null action is tantamount to making it equal to the identity arrow of a. And the same goes for the other direction. We will thus have the commutative diagram:

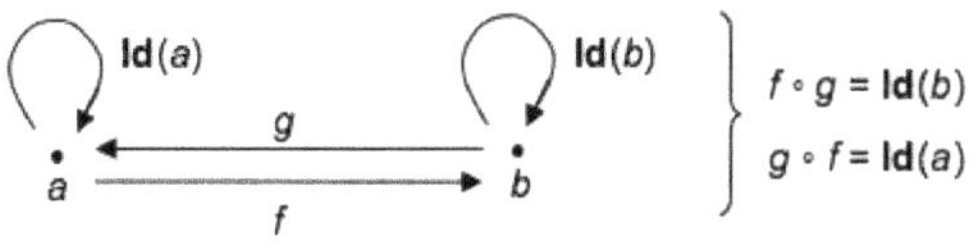

$$a \underset{f}{\overset{g}{\rightleftarrows}} b \qquad \mathbf{Id}(a),\ \mathbf{Id}(b) \qquad \left.\begin{array}{l} f \circ g = \mathbf{Id}(b) \\ g \circ f = \mathbf{Id}(a) \end{array}\right\}$$

If there is such a pair of arrows between two objects we will say that these two objects are isomorphic. The arrow f will be called an *isomorphism* from a to b, and the arrow g will be called the *inverse* arrow (obviously this is also an isomorphism, and we can also say f is the inverse of g).

Two isomorphic objects are 'literally' different, but categorially identical. Because categorial identity is specified by the network of actions of which an object is the source or the target. However, if a is in such a network, and b is isomorphic to a, b is *also* in the network. Why? Because if a is the target or the source of an arrow, and you compose this arrow with the isomorphism or its inverse, the latent 'nullity' of the supplementary action alters nothing, and consequently you might just as well write b in place of a in this type of network.

As an example: suppose that in a category we have an initial object 0 (the dual of a terminal object) and that the object b is isomorphic with 0. I say that b is *also* an initial object. Recall that an initial object is an object such that there is *a single* arrow from it toward every object of the category. There is therefore a single arrow f from 0 to b, and as b is supposedly isomorphic to 0, this arrow is an isomorphism, which admits an inverse arrow g going from b to 0. If you now take any object c of the category, there is a unique arrow h from 0 to c. The composition of g (which goes from b to 0) and of h (which goes from 0 to c) gives an arrow that goes from b to c. We still need to prove that this arrow is unique. Suppose there is another arrow, k, which will give us the following diagram:

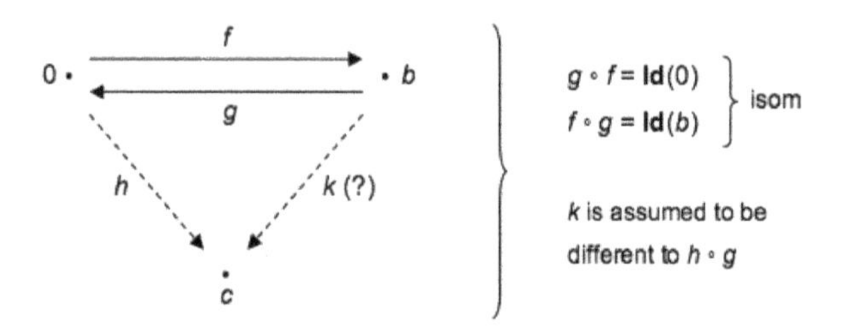

Consider in the diagram the arrow $k \circ f$. It is an arrow from 0 to c, but since 0 is initial, there is only one arrow from 0 to c, which is h, and therefore $k \circ f = h$. It follows that $(k \circ f) \circ g = h \circ g$. Thus (by associativity) $k \circ (f \circ g) = h \circ g$. However, since f is an isomorphism, of which g is the inverse, $f \circ g$ is a null action, so finally we get $k = h \circ g$. Consequently, the arrow k is indeed the same as the arrow $h \circ g$, proving the uniqueness (following the existence) of an arrow from b to object c. This clearly demonstrates that the object b, isomorphic to an initial object, is itself an initial object.

We will also demonstrate how every object isomorphic to an equalizer is an equalizer, how every object isomorphic to a pullback is a pullback, and so on.

Our fundamental concepts therefore specify some objects (when they exist in a category) as unique *up to isomorphism*. We are here in a meditation on the One. What does it mean to say there is 'one' initial object in a category?

What this actually means is:

- **a** That every object isomorphic to an initial object is itself initial. (We have just demonstrated this.)
- **b** That if there is an initial object 0, and another object, say b, which is also initial, it is isomorphic to the first.

Let's demonstrate this second point.

Take two initial objects 0 and b in a category. Since 0 is initial, there exists a unique arrow f from 0 going to b. Because b is initial there exists a unique arrow g from 0 to b. I say that the pair of arrows (f, g) defines an isomorphism between 0 and b. In fact, the composite $g \circ f$ is the arrow that goes from 0 to 0. There is indeed one such arrow, the arrow $\mathbf{Id}(0)$, and this arrow is unique *because 0 is initial*. Consequently, $g \circ f = \mathbf{Id}(0)$. We can equally show that, assuming b is initial, $f \circ g = \mathbf{Id}(b)$. These inverse compositions give us null actions, and consequently 0 and b are isomorphic.

We equally demonstrate that, in a given category, two pullbacks (products, equalizers, co-products, pushouts, co-equalizers) for the same objects or arrows are isomorphic.

Thus *the ambiguity of the One* in categorial thought: two 'different' objects (in the set-theoretic sense) can be identical insofar as the structural concepts under which they fall are the same. They will have the same conceptual *names*. Only *as letters* are they finally distinct. This suggests that the relationship signifier/letter will be thought quite differently depending on whether the underlying ontology is set-theoretic or categorial. We will say that for set theory, *every letter is virtually a name*. Whereas for category theory the reign of ambiguity through isomorphism means that some names can indifferentiate some distinct letters.

When speaking of the letter, we must always specify the ontological presentation in which we operate.

8 EXPONENTIATION

It is practically impossible to *think* the categorial definition of exponentiation without passing through its hidden set-theoretic model. Certainly we can *give* this definition, but then it would still remain a pure real, an obstacle in the way of our understanding. Which is why I hold that exponentiation is like the categorial passage where the demands of set-theoretic ontology operate, though not declared, like a forcing. Desanti calls this the 'primitive' employment of set-theoretic concepts at the heart of a theory of categories which claims to surmount it.

So let's revisit set theory. Take a set A and a set B. What is exponentiation, denoted B^A? It is, very simply, *the set of functions from A to B*. If f is a function from A to B (which takes its arguments from the set A and its values from the set B), noted $A \xrightarrow{f} B$, this function f will be an *element* of the exponential set B^A. Since in set theory a function *is itself* a set, there is no problem: the elements of B^A are the particular sets that are the functions from A to B. We write this: $f \in B^A$.

The 'generalization' that comes to mind is the following: take a category and, in this category, two objects a and b. We will call exponentiation of b by a, which we will denote b^a, *the set of arrows from source a and target b*. But 'set' is meaningless here. Our understanding of exponentiation is obstructed inasmuch as the idea of an object that would constitute a whole from a certain number of arrows is a properly set-theoretic idea. We don't even know what 'to belong to an object' means! And we don't know how to define *active* configurations, namely, diagrammatic configurations of relationships or movements. How do we translate such an openly 'collectivizing' idea (all the arrows from a to b) into a diagrammatic universality? What kind configuration can such an object really be the limit of?

The 'trick' is to analyse the set-theoretical definition in terms of functions, to break it down, to project it onto a diagram, and to see if there is a possible generalization. You will see that this relies on the set-theoretic definition which *gives* us a set B^A, resulting in a breakdown of it in terms of relations.

We begin by 'spreading out' the set B^A by an arrow which goes from this set to the set B. This is the heart of the analysis.

Take an element of B^A. This is, we have said, a function f from A to B. If we take a particular element x of A, this means that we have $f(x) = y$, where y is an element of B. We can interpret this by saying that *to the element* f *of* B^A *and to the element* x *of* A *there corresponds a determinate element of* B, *namely, the element* y. In other words, corresponding to the pair (f, x) is the element y.

But what is a pair (f, x)? It is by definition an element of the *product* of the two sets B^A and A. The product of two sets is in fact the set whose elements are all the ordered pairs of elements from the first set and then the second. We are therefore in a situation where, taking the correspondence which associates the element y to the pair (f, x) – determined by $f(x) = y$ – we have in effect defined *a function which goes from the product* $B^A \times A$ *to the set* B. This function is defined for every pair of the type (f, x), or element of $B^A \times A$, since $f(x)$ takes its values from B. We call this function the *evaluation function*.

The idea is then to show that the evaluation function, which we'll call ev, from $B^A \times A$ to B has a kind of universal position with regard to all the functions 'in the same configuration' – that is to say all the product functions of the kind $C \times A$ to B – a universality which allows us to recover the style of categorial definitions.

Take any set C and the product $C \times A$ (this is the set of pairs (z, x), where $z \in C$ and $x \in A$). Now take a function g from $C \times A$ to B. This means that g makes an element y of B correspond to a pair (z, x).

The point is that we can 'extract' from the function g, which goes from $C \times A$ to B, another function f going from A to B. To do this we simply need to *determine* a particular element z of C, and associate to the $x_1, x_2, \ldots x_n$'s of A which are paired with z, the elements of B which correspond, through g, to the pairs (z, x_1), (z, x_2) etc.

Let's consider, in effect, *all* the pairs (z, x_1), (z, x_2), … for a determinate element z of C where $x_1, x_2, \ldots$ are the elements of A. The function g associates elements of B to each of these pairs. Thus we define *a function* f *from* A *to* B, the function which, for the chosen z of the set C, makes x_1 correspond to $g(z, x_1)$, x_2 to $g(z, x_2)$, and so on. In fact, for every element x of A, we get $f(x) = g(z, x)$.

This construction, starting from the given function g from $C \times A$ to B, associates a function f from A to B (and therefore an element of B^A) to every element z of C. What this means is that *we have associated to every function* *g from $C \times A$ to B a function from C to B^A*. We write this function $\hat{g}$, and we know that we have $\hat{g}(z) = f$.

We will also remark here that for two *different* functions g_1 and g_2 between $C \times A$ and B, we obtain *different* functions $\hat{g}_1$ and $\hat{g}_2$ between C and B^A. In fact, if g_1 and g_2 are different, there is at least one pair (z, x_1) such that $g_1(z, x_1)$ is different from $g_2(z, x_1)$. But then the function $f_1(x) = g_1(z, x)$ associated to g_1 and to the element z of C is different from the function $f_2(x) = g_2(z, x)$ associated to z and to g_2, since they are different for the argument x_1. And consequently, as $\hat{g}_1(z) = f_1$ and $\hat{g}_2(z) = f_2$, then $\hat{g}_1$ and $\hat{g}_2$ are different functions from C to B^A, not having the same value in B^A for the same z in C.

Finally, there is a *bi-univocal correspondence* between the functions g from $C \times A$ to B and the functions $\hat{g}$ from C to B^A.

The situation is then the following: take an element (z, x) of the product $C \times A$. We can transform it into an element of the product $B^A \times A$ by associating it with the pair $(\hat{g}(z), x)$. Note that we do not alter the element x of A, so we proceed 'identically' to A. This element $(\hat{g}(z), x)$ is in fact a pair of the kind (f, x). And we can transform this pair into an element y of B by applying the function *ev*, meaning that to the pair (f, x), there will correspond the element y of B such that $f(x) = y$.

On the other hand, the function g associates to the pair (z, x) the element y of B, the same element, since by definition $\hat{g}(z) = f$ only provided that if $g(z, x) = y$, we have $f(x) = y$.

In other words, you finally have *a commutative diagram* whose appearance is the following (with sets to the left, with their elements to the right):

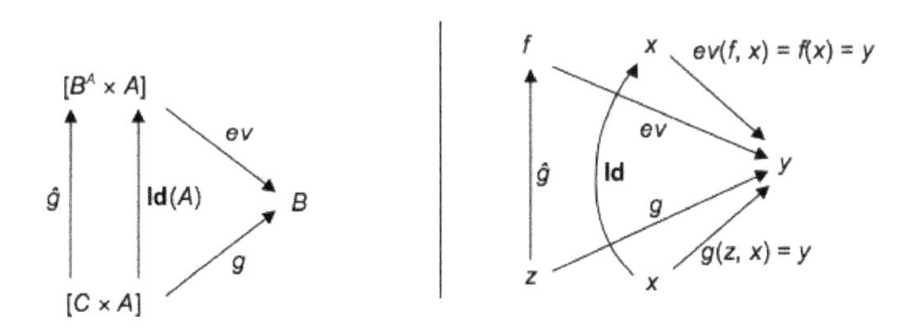

In some sense, B^A is universally positioned for the following property: 'existence of an arrow between $C \times A$ and B'. In fact, for every set C, such that there exists an arrow (here g) from $C \times A$ to B, there is a *unique* arrow $\hat{g}$ from C to B^A which makes the above diagram commute.

This is the 'extraction' of arrowed universality which analyses set-theoretic exponentiation in the categorial style.

The direct categorial projection obviously supposes that we operate in a category which admits products (see above). We will then say that the category admits exponentiation if, given two objects a and b, there always exists an objet b^a, and an arrow *ev* (evaluation arrow) from $b^a \times a$ to b such that, for every object c and every arrow g from $c \times a$ to b, there exists a unique arrow $\hat{g}$ with the following 'commutative' diagram:

Note: I placed 'commutative' in quotation marks because we have *two* arrows on the left side, one from c to b^a, and one (the identity arrow) from a to a. To restore this duality to unity, we need to know the product (or the combination) of *two arrows*.

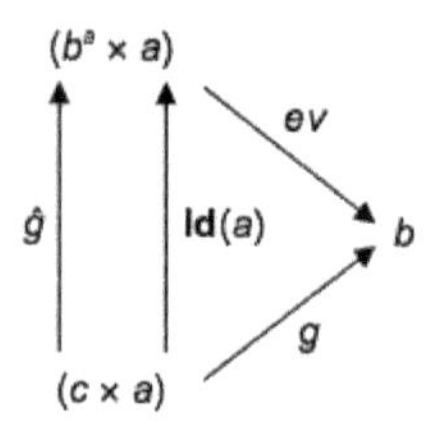

Take two arrows, f from a to b, and g from c to d. If the category admits products (as we've assumed from the beginning of this study of exponentiation), there exists the product of the two sources of f and g, i.e. $a \times c$, and of their two targets, namely, $b \times d$. We will call *product arrow* of f and g the arrow which goes from $a \times c$ to $b \times d$, an arrow whose existence and uniqueness are guaranteed by the following diagram (to understand this better refer to the diagram defining the product of two objects as limit):

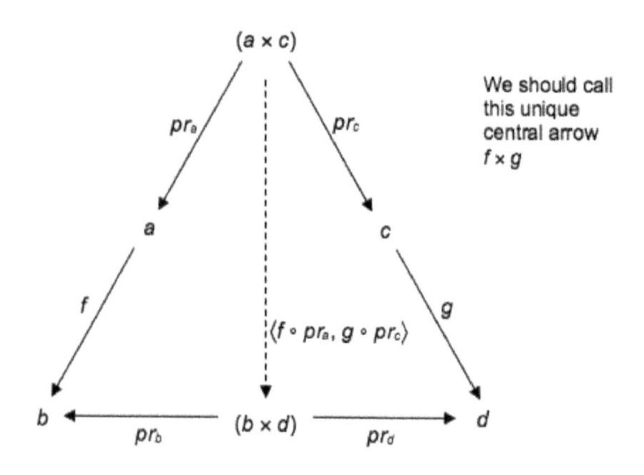

We thus reduce our duality between $\hat{g}$ and $\mathbf{Id}(a)$ to the product $\hat{g} \times \mathbf{Id}(a)$, and we define the object b^a by the following commutative diagram:

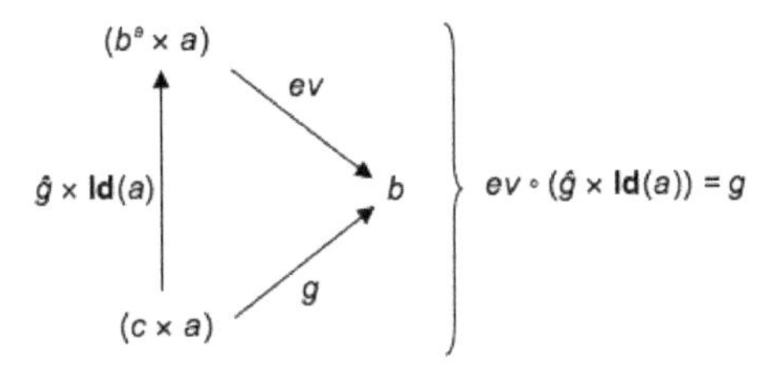

We can easily assimilate this definition by reasoning on the basis of the *empty place*. Suppose that ( ) × *a* expresses the empty property 'being produced with *a*'. We can thus draw the following quasi-diagram:

( ) × *a* *b*

A kind of limit for the diagram is: being universally positioned for the property 'being produced with *a*, and seeing *b* from this product'. This is the precise definition of the object b^a. Because:

- It is a cone, inasmuch as it is produced with *a*, and an arrow leaves from this product $b^a \times a$ to *b*, namely, the 'evaluation' arrow.
- It is a limit, insofar as, if another object *c* has the same property (an arrow *g* leaves from $c \times a$ to *b*), and is therefore itself also a cone for the diagram, there exists a unique arrow $\hat{g}$ from the second cone toward the first which makes the diagram commute.

We equally show (and this strengthens the One-dimension of the limit), as in the case of set theory, and by the same reasoning, that the correspondence between *g* and $\hat{g}$ is bi-univocal: to arrows g_1 and g_2 (which we suppose to be different) between $c \times a$ and *b* correspond different arrows $\hat{g}_1$ and $\hat{g}_2$ between *c* and b^a.

Finally, the idea of 'all the arrows from *a* to *b*' is transformed into a universal position with regard to the property 'seeing *b* from the product with a'. Whence the trace of the original set-theoretic intuition is at once perceptible and nearly effaced.

9 UNIVERSE, 1: CLOSED CARTESIAN CATEGORIES

We can now return to the idea of a sufficiently 'large' categorial universe.

The basic idea is that this is a category in which there are sufficient limits and co-limits, thus a category from which we can 'see' universally from various configurations within the category.

Indeed, a category is already a very rich universe if all its *finite* diagrams (comprising a finite number of objects and arrows) admit limits and co-limits.

Thus we propose the following definition:

A category is 'closed Cartesian' when all its finite diagrams admit a limit and a co-limit, as well as exponentiation.

In fact, it can be demonstrated that if a category possesses a terminal object, and if it admits pullbacks, then it admits a limit for every finite diagram. By duality, it follows that if it possesses an initial object, and if it admits pushouts, then it admits a co-limit for every finite diagram.

We can then say that a category is 'closed Cartesian' if it has an initial object, a terminal object, pullbacks and pushouts, and furthermore admits exponentiation.

What is Descartes' role in this affair? It is that a category must always be thought as a geometric universe. But the limits and co-limits are consistent with this geometric disposition. And this consistency, which rests finally on equations (those which govern the commutation of diagrams), can be said algebraically. In this sense, the existence of limits for every finite diagram realizes the universe as consistent with a kind of algebraic geometry. And it happens that Descartes is the inventor of algebraic geometry.

As for closure, it is taken here in its ordinary mathematical usage: operations (in this case limits) can be performed *without leaving the universe under consideration.* It is thus closed for these operations. We can also say that the universe contains a significant proportion of its own universality.

There are some very interesting properties to a closed Cartesian category, which are universal properties. I will cite one, which typifies what we mean by 'extension' of a categorial universe. Let *C* be a closed Cartesian category. We know it has an initial object, 0, and a terminal object, 1. We can also demonstrate that if 0 is isomorphic to 1, the category is degenerate. What is a degenerate category? It is a category where every object is isomorphic. Otherwise said: the difference (in the categorial sense) of 0 and 1 'supports' the general existence of difference in the category. Everything happens as if the objects of the category were unfolding 'between' the initial object 0 and the terminal object 1. We can demonstrate this theorem in stages:

a In a closed Cartesian category, $0 \times a$ is isomorphic to 0 for all objects a.

By the definition of exponentiation, we know that for any arrow g from $0 \times a$ to b (any object) there corresponds a unique arrow $\hat{g}$ from 0 to b^a. But there is only a single arrow from 0 to b^a, since 0 is an initial object. Thus, there is only a single arrow from $0 \times a$ to b, any b whatsoever. Therefore, $0 \times a$ is an initial object. Now two initial objects are isomorphic. Consequently, $0 \times a$ is isomorphic to 0.

b In a closed Cartesian category, if there exists an arrow from a to 0, then a is isomorphic to 0.

If there exists an arrow f from a to 0, you have, by the definition of the product $0 \times a$, the following commutative diagram:

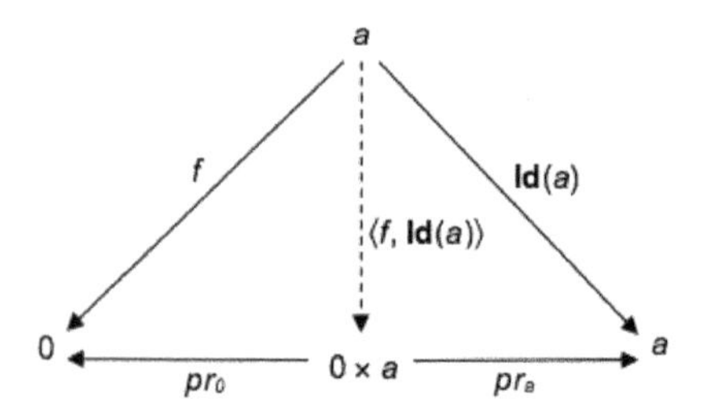

It follows that the projection pr_a is an isomorphism between $0 \times a$ and a, thus between 0 and a. In effect:

- On the one hand, $pr_a \circ \langle f, \mathbf{Id}(a) \rangle = \mathbf{Id}(a)$ [commutativity]
- On the other hand, $\langle f, \mathbf{Id}(a) \rangle \circ pr_a$ is an arrow from $0 \times a$ to $0 \times a$. Now there is only one, because (cf. above) $0 \times a$ is initial, and thus is necessarily the same arrow. Therefore $\langle f, \mathbf{Id}(a) \rangle \circ pr_a = \mathbf{Id}(0 \times a)$. Thus pr_a admits an inverse, and $0 \times a$ is isomorphic to a.

c If 0 is isomorphic to 1, the (supposedly closed Cartesian) category is degenerate.

Take any object a whatever. There exists an arrow f from a to 1, since 1 is terminal. But there exists an arrow i from 1 to 0, since 1 and 0 are isomorphic. There thus exists an arrow, $i \circ f$, from a to 0 and therefore, by consequence of the above, a is isomorphic to 0. Therefore every object a is isomorphic to 0, and the category is degenerate.

10 STRUCTURES OF IMMANENCE, 1: PHILOSOPHICAL CONSIDERATIONS

In set-theoretic ontology, the determination of the One (*this* multiple, such that it can unequivocally receive a proper name) is strictly immanent, because a set is identified by its elements, by the sets that *belong to it*. Such is the foundational character of the sign of belonging, $\in$, which is the veritable index of being *qua* being.

On the other hand, being-in, as general form of immanence (i.e. the modalities according to which one multiple can be said to be 'in' another) is radically split:

1. There is the foundational relationship of belonging, which states, for example by writing $e \in E$, that the multiple e is an *elementary* constituent of the multiple E.
2. There is the derived relation of inclusion, which states, by writing $A \subset E$, that A is a *part* or a subset of E.

Between these two relations through which being-in (or what, in the European patois, we call 'dansein')[38] is formalized there is a crucial disjunction, which is like the real of being *qua* being. The second in fact *exceeds* the first in an errant or immeasurable way. Because the *power* (the pure multiple-quantity) of the set of its parts surpasses (infinitely, but only the infinite is ontologically decisive) in a properly unnameable way the power of the set of elements, which in effect constitutes the initial set itself as One (since it *is* the recollection of its elements).

Thus a multiple's basic immanent resource (regulated by $\in$, belonging) is 'absolutely' surpassed by the immanent resource of parts, or subsets (regulated by $\subset$, inclusion).

This disjunction is at the heart of *Being and Event*, where I establish that it ontologically determines the grand forms of orientation in thought (constructivist, generic or transcendent), and, in particular, the concept of truth as infinite process passing from the undecidable to the indiscernible. This is an entirely new and paradoxical connection between the figures of 'dansein' and the concept of truth.

In the categorial formulation we find *neither* the foundational dimension of a relationship like belonging, *nor* the disjunction between two modalities of being-in.

The problem of immanence is in fact addressed through entirely different means, since *every determination is external (by arrows or relations).*

We might say that immanence is primitive in set-theoretic ontology (the sign $\in$ as 'qua' for being, inasmuch as it is such being). In the categorial formulation immanence is a result, and to some extent metaphorical.

Certainly, it is a question of progressively determining the *singularity* of an object, which at first is a simple letter. We will therefore speak *as if* we were proceeding little by little toward its immanent specification, as if we were 'filling' the letter with determinations, allowing it to function as a name. We will employ words like 'sub-object' or 'element of an element'. But this lexicon should not mislead us: the real of these determinations will *always* be thought by the external objects and arrows of which the object under consideration is the source or the target.

Thus the structures of immanence, whose originarity and violent paradox determine the entire scope of set-theoretic ontology, contrarily appear in the categorial formulation as the *extrinsic limits of the letter*, which is ultimately singularized as a quasi-name. Let's say that in category theory the intrinsic is a local saturation (of a letter) of the extrinsic.

Furthermore, the constitutive disjunction of the relation between a truth and its being, one of the finer points of set-theoretic ontology, is not found in the categorial formulation. Because we shall see that an 'element' is only a particular case of a sub-object. In fact, the categorial formulation, which delivers countless *possible* Universes, is moreover profoundly homogenous with respect to what takes the place of immanence. All this runs contrary to set-theoretic ontology, which essentially exposes *a* Universe, and which is traversed by a fault whereby predication and measurement fail.

What will be used to classify category theoretic universes is in fact *not an ontological but a logical criterion.* The great opposition is that of

Boolean Topoi and non-Boolean Topoi, itself elaborated by well-pointed and ordinary Topoi.

This will gradually lead us to the following decisive idea: the categorial formulation of mathematics does not reveal the underlying ontology, but rather *the space of possible logics*. That is to say, in set theory, logic is prescribed by ontological decision, while in category theory, ontology is 'possibly' determined by logical choice.

The debate finally concerns the old question of Aristotle, that of the relation between 'onto' and 'logy' in 'ontology'.

But we will also see that categorial formulation reveals, or clarifies, within ontology itself, those logical connections that ontology, constrained by its anticipatory decisions of thought, cannot perceive. This is the 'critical' advantage of categorial restraint, suspended between possible logics by which it treats the ontological material.

We will suggest that categorial formulation *places* ontology in an extremely open logical space, which renews its clarity within thought. Sophistry begins when we claim that the investigation of logical possibilities is *itself* an ontological decision, or – it amounts to the same – when we conclude that every decision is arbitrary (or that there is no ontology of truth).

11 STRUCTURES OF IMMANENCE, 2: SUB-OBJECT

If we want to think or determine the immanent singularity of an object, which is little more than a simple letter, it is necessary to mark out some *differences*.

Of course, a difference only leaves a mark through the action of an arrow whose target is the objet in question. And the simplest possible form of such a difference, whose object, let's say b, would be affected, is that of two parallel arrows, as in the diagram:

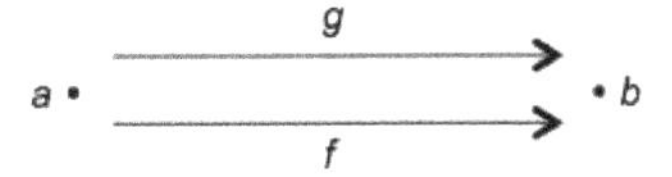

What then is a 'sub-object', if not a kind of 'internal' zone of an object where such differences are marked and 'maintained'? Yet we cannot get around the fact that such a zone must in turn be thought as the effect of an arrow on b.

We will say that a sub-object of b is specified by an arrow whose target is b such that it 'maintains' the differences induced by the parallel arrows directed toward b.

We already possess a precise concept of such an arrow: it is a monomorphism (cf. §5). Recall that this is precisely what conserves differences: if f is a monomorphism and g and h are different, then $f \circ g$ and $f \circ h$ are necessarily also different:

g

c •   • a   • b $f \circ g \neq f \circ h \rightarrow g \neq h$

h   f

As a first approximation we can therefore say that *a sub-object of b is a monomorphism whose target is b.*

Why 'a first approximation'? Because two monomorphisms whose target is b can 'affect' b – identifying its zone of conservation of differences, and therefore the sub-object – in an actively indiscernible way.

Consider the following situation: m_1 and m_2 are monomorphisms whose target is b and whose sources are a_1 and a_2. But there is an *isomorphism* (cf. §7) between a_1 and a_2, and this isomorphism makes the triangle a_1-a_2-b commute:

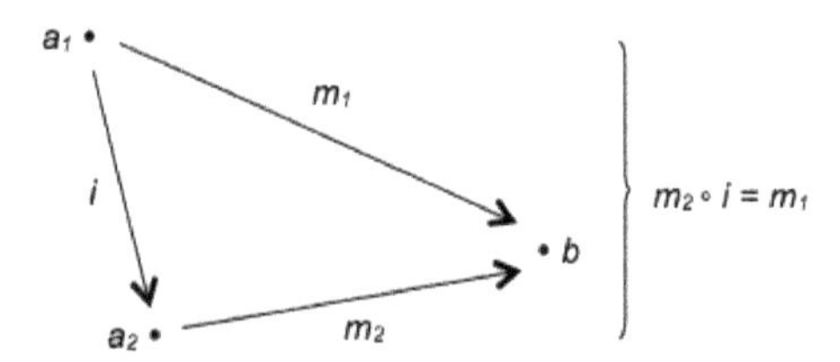

In such a case, a difference conserved by m_1 'in' b is identically conserved by m_2, and a difference that passes through m_2 passes identically through m_1.

Suppose there are two different parallel arrows that affect a_1. We have the following diagram where i and j designate the isomorphism and its inverse between a_1 and a_2:

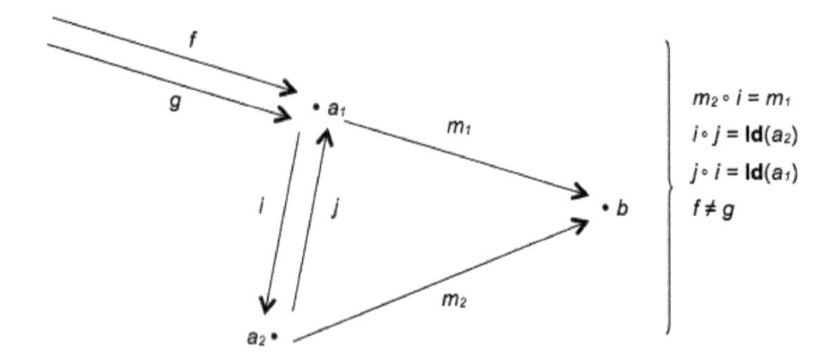

There are two different parallel actions toward b which are $m_1 \circ f$ and $m_1 \circ g$. But we also note there are two parallel actions *which pass through* m_2, namely, $m_2 \circ i \circ f$ and $m_2 \circ i \circ g$. Are they also different? Yes, because let's suppose that:

$$m_2 \circ i \circ f = m_2 \circ i \circ g$$

As the triangle commutes, we get:

$$m_2 \circ i = m_1$$

and then finally:

$$m_1 \circ f = m_1 \circ g$$

contrary to what we had previously assumed.

We could equally show (this time using j instead of i) that any difference that passes through m_2 can also pass through m_1.

In this case there is no pair of different arrows f, g which could be used to 'discern' what is marked as a 'sub-object' in b by the monomorphism m_1 from what is marked by the monomorphism m_2. We are forced to conclude that, considered as demarcating in b 'zones' of guardianship of differences, m_1 and m_2 are 'the same'.

The diagram in question can be seen as a relation between m_1 and m_2. This relation is not identity *in the sense of equality*: m_1 is indeed a monomorphism 'other' than m_2. We will note this relation $\simeq$. We will write $m_1 \simeq m_2$ every time we have the above diagram (isomorphism and commutation). And we will say that m_1 and m_2 are *similar.*

Ultimately, *a* [*un*] sub-object of b is not exactly *a* [*un*] monomorphism whose target is b. It is rather the collection of monomorphisms which are connected to m_1 (for example) by the relation $\simeq$, thus a collection of monomorphisms which are all similar to each other. In other words, it is an 'equivalence class' of monomorphisms whose target is b.

Of course, we won't hesitate to designate m_1 as a sub-object, since if m_2 is similar to m_1, it 'behaves' identically *as a sub-object of* b. But we must remember that m_1 is really the name of a *representative* of the collection, the (arbitrary) choice of one of the terms of an 'equivalence class'.

The situation is made all the more delicate by the fact that while m_1 is actually an arrow, thus a being of the category under consideration, *the same cannot be said of the equivalence class*. We have been talking about collections and classes, but these are set-theoretical terms. A set or a class of arrows is not a categorial reality. In our working definition we once again, surreptitiously, pass through set theory ontology. When choosing (set-theoretical axiom of choice!) a nominal representative of a class we return to the categorial universe by saying: m_1 is a sub-object of *b*. But 'ontologically', m_1 is the intra-categorial name of a multiple of monomorphisms, that is, an extra-categorial multiple.

We can finally say that a 'sub-object' does not designate an object, but rather an arrow. The metaphorical thought of immanence (of being-in) seeks its active outside: an action from target *b* which allows differences to pass through unaltered. A 'super-conductor' of differences connected to the object *b*.

12 STRUCTURES OF IMMANENCE, 3: ELEMENTS OF AN OBJECT

In stark contrast with set-theoretical immanentism, the notion of 'element' in categorial formulation is but one case derived from the notion (which is itself very extrinsic) of sub-object.

A sub-object of b is 'a' monomorphism (in fact, an equivalence class of monomorphisms) which marks out a zone for the conservation of differences. An element is a kind of 'minimal' sub-object, a sub-object that *only passes on that which is unique.*

How is this possible? Recall that a terminal object, properly noted 1 (cf. §5.A), is such that there is, from every object a to 1, one and only one arrow. It follows:

- **a** That every arrow from 1 to b is a monomorphism. Indeed, we cannot have two parallel arrows toward 1 from, say, an object a, because there is only one arrow from a to 1. Consequently, the arrow from 1 to b conserves differences, simply because it never admits any!
- **b** That this monomorphism, as we have just seen, doesn't mark any zone of difference in b since what it 'passes on' can only ever be unique. It is therefore just right for expressing in categorial language *the marking of elementary unity in an object.*
- **c** That if two arrows from 1 to b are similar, *they are identical.* In this case, let us consider the diagram of similarity:

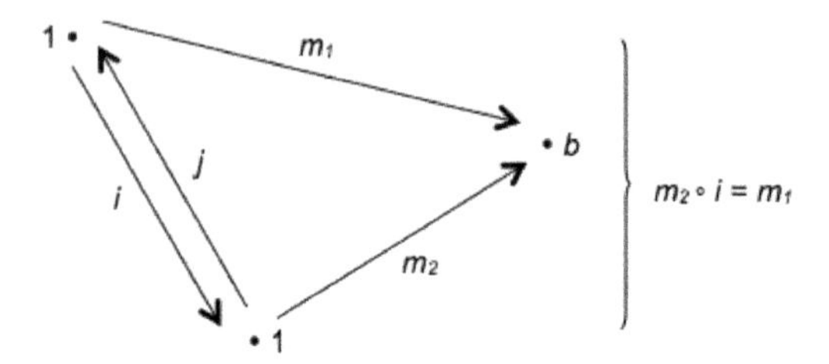

The arrow i (resp. to the arrow j) goes from 1 to 1, thus it is unique (since 1 is terminal). Yet there is necessarily another arrow which goes from 1 to 1, namely, **Id**(1). Consequently, i = **Id**(1). And as the diagram commutes, we have $m_1 \circ \mathbf{Id}(1) = m_2$, which is to say $m_1 = m_2$ (nullity of the identity action).

This means that the source object 1 unequivocally marks b with the sign of the one (if there is an arrow from 1 to b) without 'drawing' equivocations of similarity with itself: to two *different* arrows going from 1 to b correspond two effectively different (and not possibly similar) sub-objects.

For all these reasons, we will call *element* of an object b any arrow from 1 to b. This is to say any sub-object whose source is a terminal object. It is in this way that the elementary marking of the one in an object in a category is carried out.

Does it follow from this that we can in some way 'identify' an object b by its elements, by the arrows from 1 to b, as a set is identified by the elements immanent to it? Not at all! For starters, an object of a category is in no way required to have even a single element! This is the categorial definition of the void: *an object is void if it has no element.* Formally, this definition is the same as that of the void set in set theory. The big difference is that in set theory ontology the 'void' singles out *a* set, which is the point of departure for any thought of the pure multiple, which is like the suture of thought to being as such (presentation of unpresentation). Whereas in numerous categories (or possible universes), all kinds of different objects can be considered perfectly empty: it is enough that no arrow goes from 1 toward them. They are therefore subtracted from any elementary marking by the One.

There is one object that we can be assured is empty: this is the *initial* object, noted 0, at least if the category is closed Cartesian (cf. §9) and not degenerate (idem: a category is degenerate if all its objects are isomorphic, which is to say that it only has one object, in the sense that identity, in category theory, is always 'up to isomorphism'). We have in fact shown (idem) that if, in a closed Cartesian category, there is an arrow from 1 to 0, then the category is degenerate. *A contrario*, even if there are in a category of this type only two really different (non-isomorphic) objects, then it is certain that there is no arrow from 1 to 0, which is to say that 0 has no element, that it is empty.

Is it the case, reciprocally, that every empty object is in fact a 0, that is to say, an initial object? This would bring us back somewhat to the uniqueness (up to isomorphism) of the void, since two initial objects, two 'zeros', are isomorphic. This would therefore bring us closer to set theory, where the void is unique, and coincides with the 'origin' of the hierarchy of multiples.

This question is very profound, touching on the thought of being. We'll return to it presently. It suffices for the time being to say that the response is, in general, negative: in the majority of possible categorial universes, there are void objects that are not initial, and 'void' does not coincide with 'zero'. The categorial formulation renders the void itself equivocal.

13 'ELEMENTARY' CLARIFICATION OF EXPONENTIATION

We began our (painful) exposition of exponentiation contending that, in set theory, a function from A to B is an *element* of the set B to the power of A, namely, B^A.

Do we find, in categorial exponentiation, any sign of this construction 'by membership' of B^A? We can pose the question now that we possess a categorial concept of 'element' (arrow whose source is 1). In other words, can we examine more closely, thanks to the concept of element, the idea that an arrow from a to b is an 'element' of the set b^a?

This sign can moreover be found *in the form of a sign*. In fact, an element of b^a (thus an arrow from 1 to b^a) is associated to every arrow f from a to b. By a profound intuition, categoricians have called this associative element the *name* of the arrow in question.

So we can say for every arrow from a to b, not that it *is* an element of b^a, but that it is *named* by an element of b^a.

Nothing is more characteristic of the divergence of styles between set-theoretical ontology and the categorial formulation of what is possible for being than that marked by the shift from belonging (the function f *is* 'insofar as' it is an element of B^A) to nominal representation (the arrow f is named by an element of b^a).

To get this 'elementary' name we first consider the product $1 \times a$. We know that there is an arrow that goes from $1 \times a$ to a, which we have called the 'projection' of a, or pr_a (cf. our study of the product in §5B). We also assume that there is an arrow f from a to b. Hence the composite $f \circ pr_a$ is an arrow from $1 \times a$ to b. We are under the conditions of the schema of exponentiation, because $b^a \times a$ (and the arrow ev to b) must be universally positioned with regard to $1 \times a$ as well as for the arrow described above going from $1 \times a$ to b. In particular, this arrow must have a corresponding arrow from 1 to b^a, much like our g corresponded to $\hat{g}$ (see §8). It is this arrow (as in the case of $\hat{g}$ for the situation already described) which will be the name of f, and it is indeed an element of b^a. We will write $n(f)$ for the 'name of f. See the diagram:

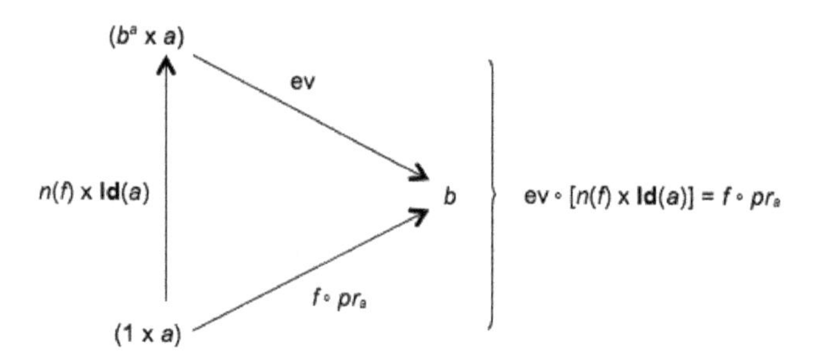

This construction is typical of the categorician's skill when it comes to taking up a prescription of set-theoretical ontology and 'spreading it out' equivocally: what was immanently given by a function as an element of a set here becomes nomination of an arrow by an element (itself extrinsic) of an object.

And things don't end here: in set theory, the 'evaluation' function is that which makes an element f of B^A and an element x of A correspond to an element $f(x)$ in B. The function *ev* is simply a recapitulation of any function from A to B (thus any element of B^A) taking its argument in A and its values in B.

What becomes of this simplicity in the categorial apparatus? We start by supposing an element of a, say x, considered this time as an arrow from 1 to a. Note we are in no way obliged to think that such an element exists, because a could in fact be empty… But let's assume anyway that it is not, and that there exists an element $1 \dashrightarrow a$. So we have established that *ev*, applied to the product-arrow of 'name of f and the element x, or $ev \circ \langle n(f), x\rangle$, gives the same thing as the composition of f and x! Thus not only is every arrow f from a to b represented by a name which is an element of b^a, but moreover the evaluation of the product of the name and the element x gives us the composition of f and x – which is like a dim evocation of the 'set-theoretical' formula $ev(f, x) = f(x)$.

The demonstration of this point, which binds together the elementary name of an arrow and a (supposed) element from the source of this arrow is an excellent exercise. I will give only the initial diagram:

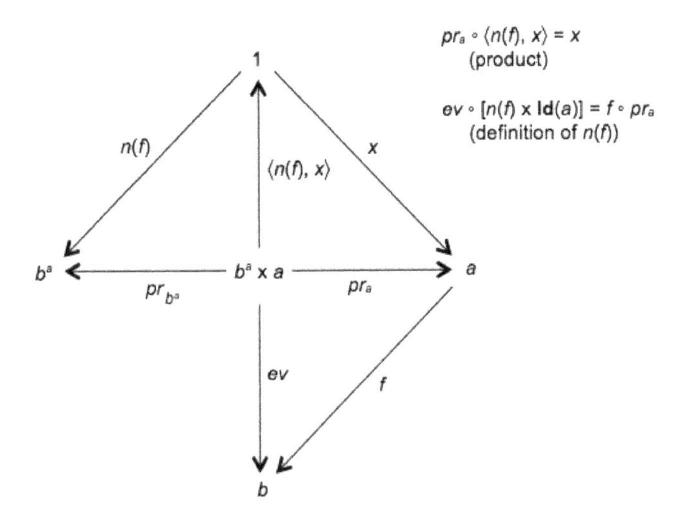

14 CENTRAL OBJECT (OR SUB-OBJECT CLASSIFIER)

Until now, the construction of possible universes (or the formal inspection of the understanding of the God of Leibniz, who actually contains every possible universe) has only given us – with, for example, the concept of closed Cartesian categories – some indications of 'dimension' (existence of numerous limits). Furthermore, the immanent investigation of objects has only delivered the general concept of sub-object, which is applicable to any object, and the subordinate concept of element, which concerns the object-targets of an action that has its source in 1. Our universes are devoid of any centre, of any internal reference that would give some measure of order to the chaotic network of actions and the ambiguous dispersion of objects.

We can also say that while our categorial universes are powerful spaces of possibility, they appear unable to contain the resources of a logic by which actions and compositions, the limits of literality through the saturation of relations, would find themselves somewhat *centred*, and thereby put to the test of an immanent logical consistency, or even an evaluation, by the universe, of the statements which are admissible therein.

It is the concept of such a centre that concludes the procedure of construction, giving us the concept of Topos as the form of every universe.

The idea of a 'centre' is not easy to conceive in the extension (which is itself unnamed, since there is no indication yet of what a categorial concept of the infinite might be) of the space of possible universes.

A first indication would be to consider a Central Object as that which is visible from any other object of the category (recall our geometrical and optical images). Let's call this object C (for central). Here there would be at least one arrow (though not necessarily only one) from any object a going to C.

But this is inert and too much like a terminal object.

The fundamental idea is that the Central Object – and it alone – *makes the connection between the 'limit' configurations, which concern the composition of arrows (commutation), and the structure of quasi-immanence given by monomorphisms*. This is why one usually calls the Central Object a sub-object classifier: by utilizing certain limit-procedures, this Object (itself specified by the play of arrows) in effect allows, as we shall see, the sub-objects to be centred, and furthermore reduces the equivocity of similarity to the clarity of equality. Thus we simultaneously attain a topical order (centration) as well as a logical order (analysis of similarity as equality).

From now on, unless explicitly stated to the contrary, we will assume that the categories we are examining are closed Cartesian (cf. §9).

A Central Object C is an object which has an element, $1 \dashrightarrow C$, called the true and noted T (here begins categorial logicization), and which is such that, for any monomorphism $a \overset{f}{\dashrightarrow} b$, a unique arrow exists from b to C, called the centration of f, and written $c(f)$, such that the diagram below is a pullback:

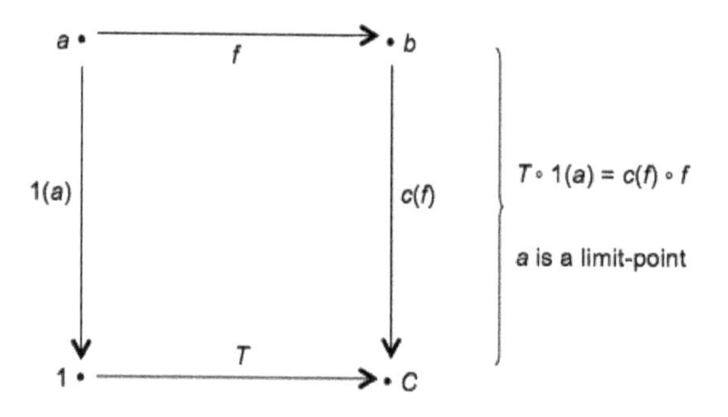

We can already see that a Central Object associates an arrow toward C to any monomorphism, regardless of its source and target. This is our program of centration, our 'topical' program.

What is the logical program? It is a theorem, the crucial theorem of the Centre:

If two monomorphisms with the same target are similar, therefore if they belong to the same equivalence class and thereby name the same sub-object, then their centration by C is the same (in the strong sense, the sense of equality). Reciprocally, if two monomorphisms have the same centration, they are similar.

In other words, we have:

$$f \simeq g \leftrightarrow c(f) = c(g)$$

Thus centration in fact reduces the concept of sub-object, caught in the set-theoretical ambiguity of the representative of a class, to the logical precision of strict equality.

Let's prove the theorem.

a Direct proposition. Suppose that $f \simeq g$ (f from a_1 to b, g from a_2 to b). There is thus an isomorphism i between a_1 and a_2, which commutes the triangle a_1–a_2–b. Let's apply this situation to the pullback of the Central Object appropriate to the monomorphism f:

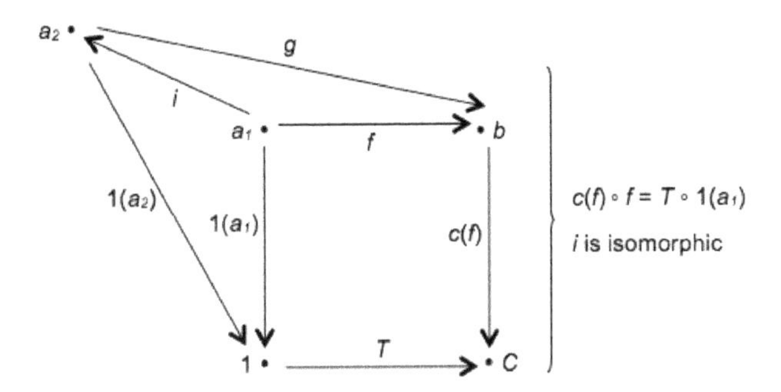

Since a_1 and a_2 are monomorphisms, and a_1 is in the position of pullback object for the square, the result (cf. considerations on the categorial indiscernibility of isomorphic objects, §7) is that a_2 is also in a pullback position, meaning that the 'square' a_2–b–C–1 is a pullback. But from the definition of C it follows in this case that the centration of f, namely $c(f)$, or the right side of the 'square', is also the centration of g, at the top, since $c(g)$ is rightly the *unique* arrow which, under these conditions, is the pullback. Thus, $c(f) = c(g)$.

b Reciprocal proposition. Suppose f and g have the same centration. In the two diagrams below, the result is obtained by interchanging f and g from the fact that they are both positioned at the top of the pullback square whose right side is the centration $c(f) = c(g)$:

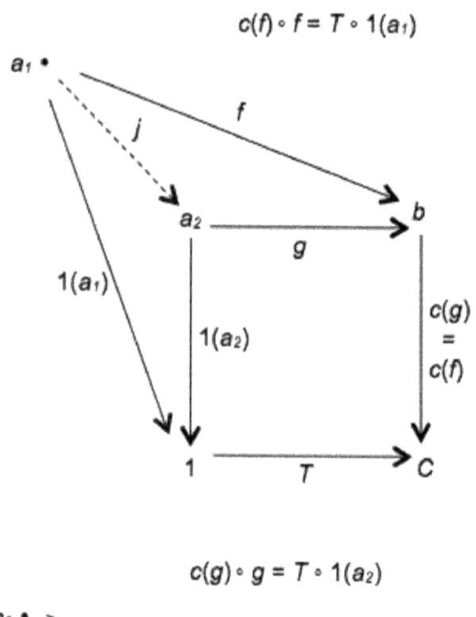

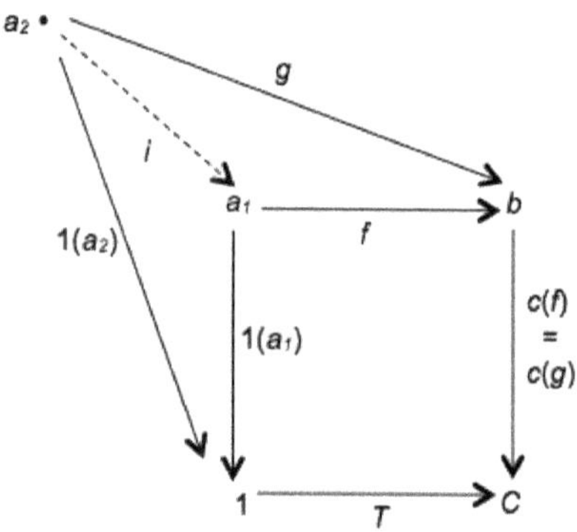

> The arrows i and j resulting from this 'exterior' commutation force the existence of a unique arrow (from a_1 to a_2, then from a_2 to a_1) which makes the entire diagram commute (this is the expression of the fact that a pullback is a limit). Yet this pair is an isomorphism between a_1 and a_2. In fact we have $f \circ i = g$ and $g \circ j = f$, hence, by substitution, $f \circ (i \circ j) = f$, and $g \circ (j \circ i) = g$, resulting in $i \circ j = \mathbf{Id}(a_2)$ and $j \circ i = \mathbf{Id}(a_1)$. So j is the inverse of i, and i is an isomorphism. As this isomorphism commutes the triangular diagram a_1–a_2–b, the result is that f and g are similar monomorphisms.

Thus we once again have the general equation:

$$f \simeq g \leftrightarrow c(f) = c(g)$$

This establishes that to two categorially different (dis-similar) sub-objects correspond two intrinsically different (non-equal) centrations.

Moreover, every centration (thus any arrow from any object toward C) corresponds to one and only one monomorphism. In fact, in a closed Cartesian category (and we assume we have one here), the pullback from T and from an arrow of target C *always exists*, because it is the limit of the finite diagram:

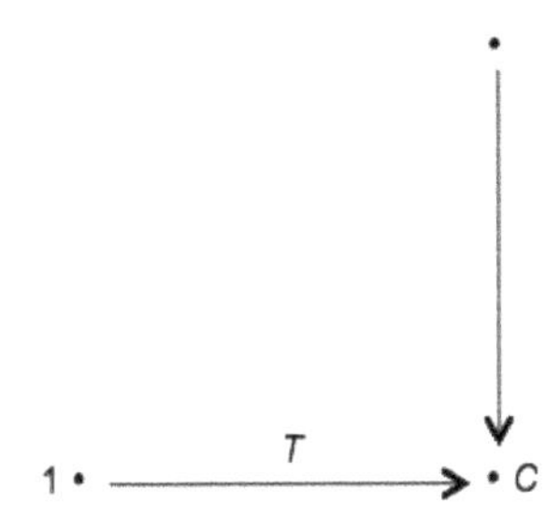

Let's assume the completed pullback:

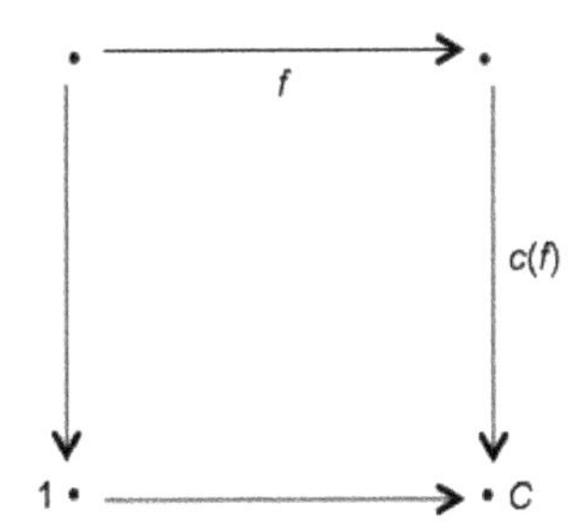

The arrow f which appears at the top:

- is a monomorphism (this is the result of a basic lemma which we'll demonstrate shortly);
- necessarily has as its centration the arrow on the right side

Finally, we obtain (in set-theoretical language!) a bi-univocal correspondence between the family of distinct sub-objects (effectively dis-similar monomorphisms) and the family of arrows of which the centre C is the target. In other words, every sub-object has a centration (an arrow toward

C), two different sub-objects have different centrations, and every arrow toward C is the centration of a sub-object.

But we are still far from having exhausted all the resources of the Central Object.

Note 1. The attentive reader will remark that I sometimes say 'the central object' and sometimes 'a central object'. Are there, in the end, one or many central objects in a category? The answer is exemplary of category theory: if there are many central objects, they are isomorphic!

Note 2. For speculative reasons, I have chosen not to use here the exact vocabulary of the mathematical literature. As I have said, what one generally calls 'sub-object classifier' and notes Ω is what I here call Central Object, or Centre, and note C. Likewise, one calls 'character of f, and notes X_f, what I call centration of f and note $c(f)$.

15 THE TRUE, THE FALSE, NEGATION AND MORE

The reason why the element of C which serves as the basic machinery for the centration of sub-objects is called 'true' will only become clear little by little. This reason engages the rebellious relation that all thinking establishes between the unpresentation of being and the procedure of truth.

It requires some remarks, all the same.

a What is the centration of the true?

The true must have a centration, since an arrow of the kind $1 \dashrightarrow C$ is a monomorphism (see the remark in §13 above). It suffices, as it so often does in category theory, to look at the following diagram:

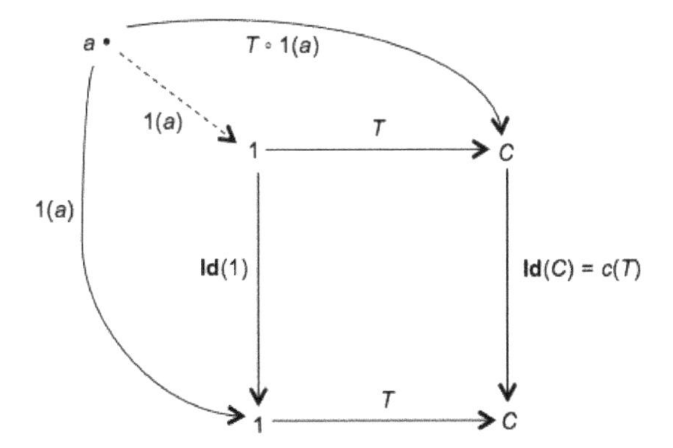

It appears that the square is a pullback, and as its base is precisely T, it is a 'central' pullback. Hence *the centration of the true, $c(T)$, is the identity of the Centre,* **Id**(C).

Doubtless this is the naked expression of that originary relation, thought by Parmenides, between truth and identity. Let's say that when centred, the true effectively presents itself as a null action, as a pure identity. We will add however that a truth is something entirely other than a centration of the true, which constitutes precisely the maxim of knowledge.

b What is the true the centration of?

Again there must be a monomorphism whose centration is the true, because every arrow of target C is the centration of some sub-object (cf. end of the last section).

Let us look, as we simply need to see, clearly and distinctly:

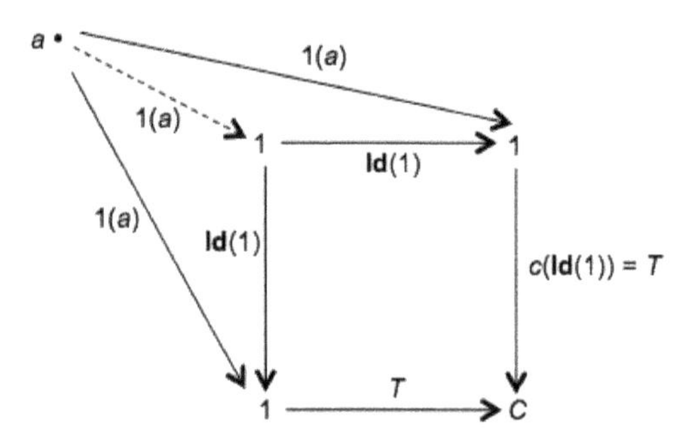

The conclusion is obvious: *the true is the centration of the identity of 1, the terminal object. Thus* $c[\mathbf{Id}(1)] = T$.

This time, it is the correlation between the true and the *extremum* of identity that is at stake. You get the true by focusing on 'elementary' identity, that which fixes the One to the horizon of visibility of any universe. We should add that the true delivered as the result of an elementary centration, or in short as the centration of the power of the One, shouldn't be confused with a truth.

c And the false?

Since an element of the Centre is named the true, where is the false? Is it unreasonable for it to also be an element of the Centre, indeed the *other* element of the Centre? These questions will lead us far, to the point where, from the interior of the possible, classical and non-classical logics, the analytic and the dialectic, confront each other.

Let's say for now that *the false is the centration of the (unique) arrow which goes from the initial object 0 toward the terminal object 1.*

Here then is the pullback constitutive of the false.

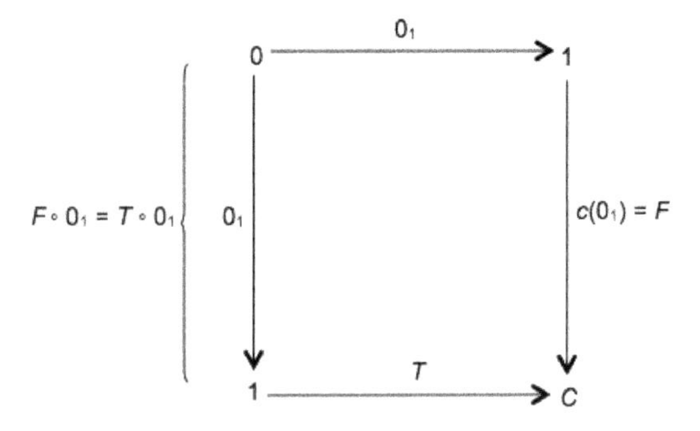

It is obvious that the false, marked *F*, is an element of the Central Object. This Centre is decidedly not empty (on the void cf. §13). Here, it has two elements, the true and the false.

But are they really two? We should be wary: in the categorial universe, difference is cunning, and identity evasive. The true and the false are after all two arrows, two monomorphisms. Moreover these elementary monomorphisms have the same source (1) and the same target (*C*). Can they not be, 'these' arrows, two names for the same act? We would then have a kind of rational scepticism, where truth-values superimpose (as in the thought of Nietzsche) their nominal duality upon an identical principle of power.

This is the occasion to give the only *demonstration* that I know of the following strong philosophical statement: *if the universe is not degenerate, that is to say if it contains real difference, then the true and the false cannot be, as actions or arrows, identical.* Or in short: the true is intrinsically different to the false, since the Universe is multiple.

Thus, the existence of difference in general induces the true and the false. The multiple as figure of being (here the multiple of objects) requires the fundamental logical multiplicity (at the very least, duality): truth-values.

Suppose that the universe is indeed multiple. This is to say: the category is not degenerate. If the true and the false were identical, the sub-objects of which they are the centration would be similar, by virtue of the fundamental property of the Centre, which changes the similarity of sub-objects through the identity of their centrations.

Yet we have just seen that the true is the centration of **Id**(1), while the false is the centration of 0(1), the (unique) arrow which, traversing the whole of the category, goes from the initial to the terminal object. If these two arrows were similar, we would then say that their sources are isomorphic, and therefore that 0 is isomorphic to 1. We have known for a while now that if 0 is isomorphic to 1, the category is degenerate (cf. §9.). However, we have just claimed that it is not. Consequently, the true is distinct from the false.

But could there not be between them, since their identity is impossible, a relation of similarity? This time it would be a tempered scepticism, which allows the distinction between truth-values, on the grounds of their essential resemblance. The answer must be no, because *T* and *F* are *elements* of *C*. And we have shown (cf. §13) that two different elements, like *T* and *F*, cannot be similar.

It is therefore certain that if the universe affirms even a single difference, it is necessary, on the basis of its Centre, firstly that true and false exist as arrow-elements of the Centre, and secondly that they are neither identical nor similar, but themselves really different.

d Another question: does not the true knowledge of the difference between the true and the false require negation? The affirmation of the duality true/false as elementary arrows of the Centre does not give us the concept of their relationship. What can we say of negation in a categorial universe?

The response, moreover, is at the same time very simple and very profound: *Negation is the centration of the false.* This gives us the rare pleasure of seeing negation, which we mark as ∾:

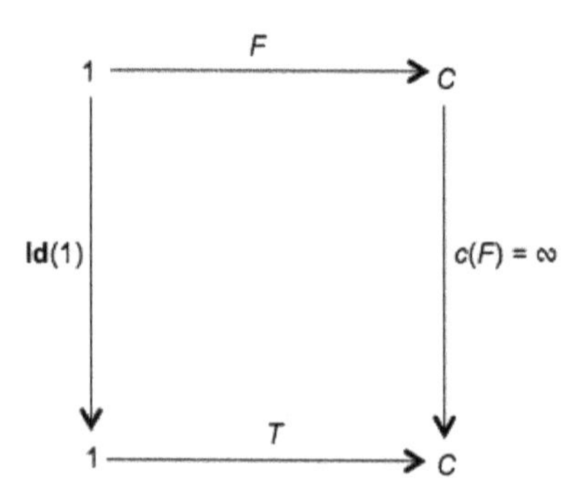

Thus negation is itself an arrow, and therefore an action *internal to the universe*, a particular operator entirely identifiable by the machinery of the Centre.

We have already noted above that the centration of the true was the identity of the Centre. Symmetrically opposed to this Parmenidean identity of 'centred' truth is negation, which is also an arrow from C to C, but one which is as different from identity as the true is from the false. This time one could speak as a Heraclitean, opposing the centration of the true as the identity of the Centre (thus a fixed point of the universe) to the centration of the false as errant activity, mobile power that puts C to work against its own identity.

e Last question: how does the negation arrow *operate*? Specifically, does it determine the ratio between the true and the false?

First notice that the true, the false and negation are all arrows. It is a striking particularity of categorial formulation that logical operators and the 'values' attributable to statements are all presented *on the same level*. This point will definitively shape our thinking of what is thought in Categories. Moreover, this unicity of levels is 'active': negation and values are actions, relations, or 'perspectives'. In particular, negation acts in combination with other arrows, which is not surprising, but it is the same for the true and for the false, which, as Nietzsche would rejoice, are initially active affirmations or powers, and not Ideas. For example, it is perfectly legitimate to write:

$$∾ \circ T \circ \mathbf{Id}(1)$$

$$F \circ 1(c) \circ T$$

$$F \circ 1(c) \circ ∾ \circ \mathbf{Id}(c)$$

which combine the *actions* of the true, of identity, of negation, and of the false, to other actions which crisscross and make up the universe.

Now consider the following diagram:

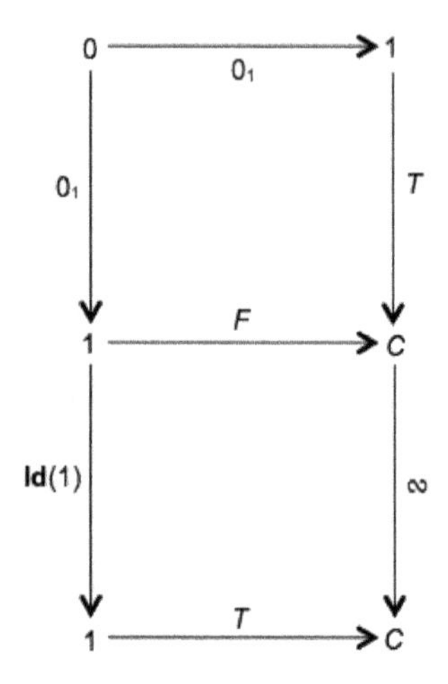

The bottom square is the pullback which defines the negation as the centration of the false. It is thus commutative, and in particular:

$$\infty \circ F = T \circ \mathbf{Id}(1) = T$$

It is thereby established that in categories of this type (closed Cartesian with a Central Object), *the action of the true is equal to the composition of negation and the False.*

The square at the top is also a pullback, which defines the false as the centration of the arrow 0···➤1.

Note: This calls for consideration: compared to its *canonical* presentation, the pullback in this diagram *is turned around*: first turned counterclockwise around point 1 on the bottom left, then flipped around the bottom arrow 0···➤1, movements we can follow thus:

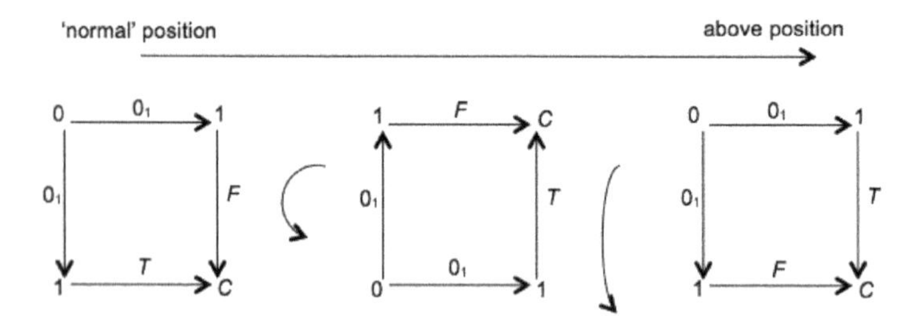

But of course, the only thing that counts in a diagram is the arrangement of the composition of arrows, and these movements are in no way altered by this arrangement. It is an exercise for the eye following the connections that remain invariant in these apparently very different diagrams. If we call geometry the study of the aspect of perceptible deformations in space that remains invariant, then the diagrammatic presentation of these combinations of arrows is indeed a geometric exercise.

We can show that if two squares placed one on top of the other, as in our initial diagram, are pullbacks, then the perimeter of the larger rectangle this unit forms is also a pullback. This is called 'the pullback lemma'.

The pullback lemma, or PL, thus shows us that the following diagram is a pullback (this is the 'squared' rectangle):

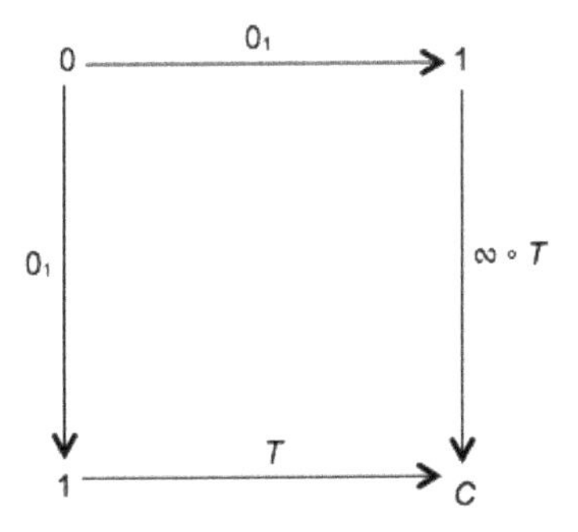

We can recognize a pullback of centration (since the arrow at the bottom is T, and the arrow on top is certainly a monomorphism). Consequently, the right side is the centration of this top arrow, which means that the negation of the true is the centration of 0···►1. Yet, the 'centration 0···►1' is the very definition of the false. And two arrows which have the same centration are similar (fundamental theory of the Centre). But two similar *elements* are identical (cf. §13). Now, the source of F and $\infty \circ T$ is 1, and therefore they are elements. And $F = \infty \circ T$.

Thus, just as the true is the composition of negation and the false, so too the false is the composition of negation and the true. Negation, as a categorial arrow, acts in conformity with its usual logical signification.

We already see the sense in which a categorial universe *presents a logic in an immanent way*, through actions which are *part of* the universe.

16 THE CENTRAL OBJECT AS LINGUISTIC POWER

In set theory ontology, the decisive knot between the question of language and the question of multiple-being is given in Zermelo's axiom (or the axiom of separation). If you have a statement $P(x)$, which 'says' something about the entity x, whose attribute is, say, the property P, and if you have a multiple A, the axiom states *there exists a subset of A made up of all the elements which possess the property P*. In other words, any statement $P(x)$ 'separates' in A a *subset*, of which all the elements held in common have the property expressed by P. Thus, with regard to the multiple whose existence is affirmed, language has the power to discern a subset, or a part.

This can also be said as: given A, there exists a part B_P of A such that $x \in B_P$ means that $P(x)$ is *true* (the element x of A really possesses the property P).

Has our categorial truth, the elementary action whose target is C, the power to tie together elements, parts, and statements of the language?

A 'part', in the categorial formulation, is a sub-object, a quasi-immanent structure. An element is an elementary sub-object, therefore an element of source 1. A statement can hardly be anything other than an arrow *offered up for evaluation*. However our only principle of evaluation is the Central Object. We should say that a property P is an arrow directed toward the Centre.

Aligning a property with a part thus requires tying this property to a sub-object, therefore *producing the sub-object of which the property P is the centration*. The fundamental situation which expresses that the property P singularizes in an object b a sub-object $f_{P,}$ can then be expressed through the pullback seen here:

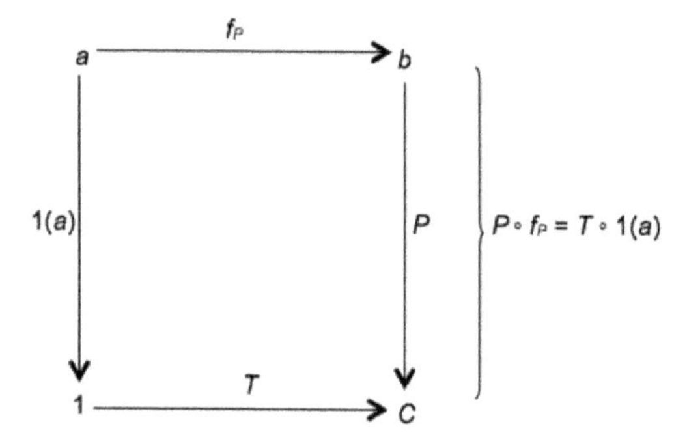

Yet nothing therein makes apparent the idea of an element with the property P.

This is because, as the sub-object is an arrow, we must, to continue the analogy with Zermelo's axiom, find out how an element can indeed 'belong' *to an arrow*.

We will define this relation with a commutative diagram. Say x is an element of b (thus an arrow $1 \xrightarrow{x} b$) and f is a sub-object of b (thus a monomorphism of target b, and, say, source a). We will say that x *belongs* to f, and write $x \in f$, if there is an element y of a (the source of f) which commutes the diagram:

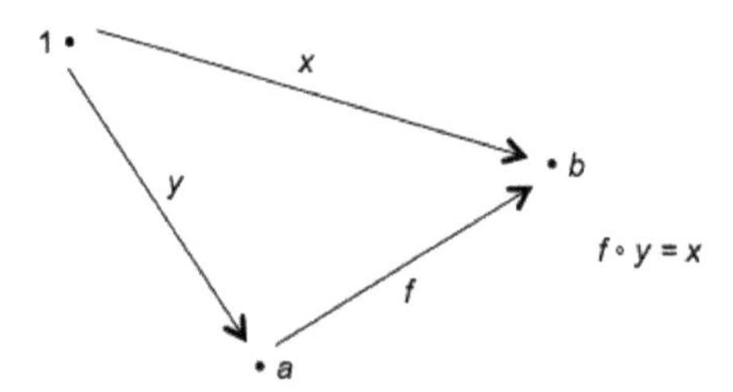

Basically, the element x belongs to the arrow f if the latter, when applied to an element, gives the element x. We are very close to the functional expression of set theory $f(y) = x$.

Note. In all this we must never lose sight of the fact that x, f, and y *are arrows*, and that the 'membership' of $x \in f$ only marks a commutative equation, namely, $f \circ y = x$, the singularity of this equation being that y and x are elementary arrows, therefore arrows whose source is 1.

Let's now suppose that there exists an element x of b. We would like to know what it means to say that x 'belongs' to the arrow f_P, which separates in the object b the sub-object whose centration is the property P. To do this there must be an element $1 \xrightarrow{y} a$ which is such that $f_P \circ y = x$. We have the following diagrammatic situation:

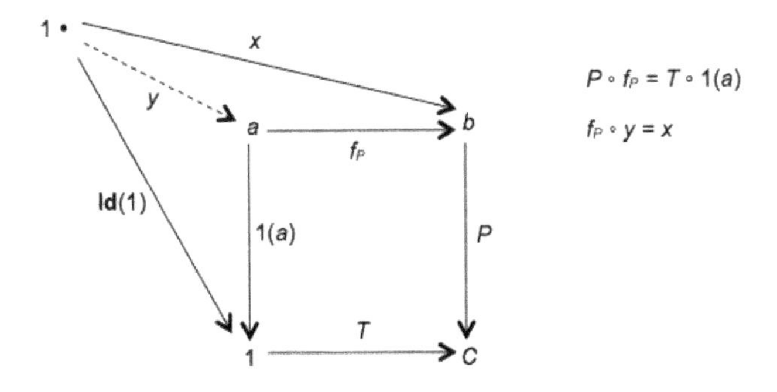

Let's note that the triangle 1–a–1 on the left necessarily commutes (because $1(a) \circ y$ is an arrow from 1 to 1, thus necessarily equal to **Id**(1)). And we have supposed that $x \in f_P$, thus that the triangle on top commutes. Yet the interior square is a pullback of centration. The arrow here therefore only exists (and is in this way unique), if the entire diagram commutes, thus if its exterior perimeter commutes (limit position of the pullback). That is, if we have:

$$P \circ x = T \circ \mathbf{Id}(1) = T.$$

Finally, we have the following statement: the element x only belongs to f_P if $P \circ x = T$.

This 'expression' is clearly equivalent to Zermelo's axiom: if f_P is conceived as the sub-object of b which in this case marks the property P, the element x of b only belongs to this sub-object if its composition with P is true.

Therefore, in the network of combined actions, the linguistic power of the Central Object is clarified.

17 UNIVERSE, 2: THE CONCEPT OF TOPOS

We call *Topos* a closed Cartesian category possessing a Central Object.

In other words, a Topos is a category:

- Where there are limits and co-limits for every finite diagram.
- Essentially, this means that there is a terminal object 1, an initial object 0, products $a \times b$ and co-products $a + b$ for every pair of objects (a, b), pullbacks and pushouts for pairs of arrows with the same target (pullback) or the same source (pushout), and an equalizer and co-equalizer for every pair of parallel arrows.
- Where, for every pair of objects (a, b), there exists a power object b^a together with its evaluation arrow *ev*.
- Where there is a Central Object C and its element T (arrow from 1 to C). But also, if the Topos is not degenerate, the false arrow and the negation.

A Topos is a possible mathematical universe, which is both 'big' (existence of limits) and centred, and which presents its own internal logic.

A non-degenerate Topos immediately presents three distinct objects (there are no isomorphs between them): 0, 1, and C. Indeed, C cannot be 0 (because C has elements, for example the true and the false, while 0 is empty, as we have shown). And C can't be 1, since there are at least two different arrows from 1 to C (the true and the false, and we have seen that these are different), yet if C was a terminal object, there could only be one arrow from 1 to C.

Let us, by way of recapitulation, 'tinker' with these initial three objects.

a Arrows

Which are mandatory? The three identity arrows of course: **Id**(0), **Id**(1) and **Id**(C). The unique arrows from 0 to C and from 0 to 1, which show that 0 is initial. The unique arrows from C to 1 and from 0 to 1, showing that 1 is terminal. The true arrow from 1 to C, and the false arrow, also going from 1 to C. And the negation (centration of the false), being an arrow from C to C which is different to the identity arrow.

b Products

We have demonstrated (§9) that for any object a in a closed Cartesian category, and therefore *a fortiori* in a Topos, $a \times 0$ is isomorphic to zero. The products of zero therefore give us no new object in the categorial sense (despite being 'literally' new objects).

What can we say of the products $a \times 1$? The die is cast by the following commutative diagram (cf. the definition of product):

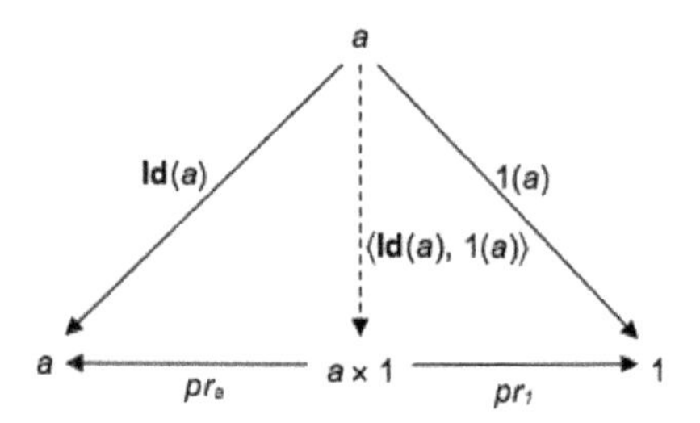

We can see that:

$$pr_a \circ \langle \mathbf{Id}(a), 1(a) \rangle = \mathbf{Id}(a)$$

and

$$\langle \mathbf{Id}(a), 1(a) \rangle \circ pr_a = \mathbf{Id}(a \times 1)$$

There is therefore a reversible arrow between a and $a \times 1$, and consequently the objects a and $a \times 1$ are isomorphs. Still, the products of 1 do not give us any 'new' object.

We finally have:

$$C \times 0 = 1 \times 0 = 0$$

and

$C \times 1 = C$, much like $1 \times 1 = 1$

It is only – in this 'local' investigation – the product $C \times C$ which gives us a new object.

From this product arises a particularly interesting situation, which is given in the following commutative diagram:

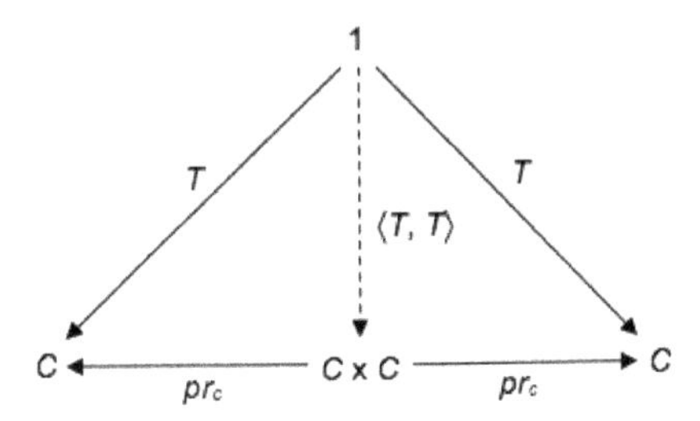

The arrow $\langle T, T \rangle$, which is an arrow from 1 to $C \times C$ (therefore an element of $C \times C$), and which exists in every Topos, serves as the point of departure for the definition (after negation) of another logical arrow. In fact we call *conjunction*, and note this ⋒, *the centration of the arrow* $\langle T, T \rangle$. This arrow expresses in a Topos the action of the logical connective 'and', as in 'p and q'. We know that this operator takes the value 'true' only if the statements p and q are at the same time valued true. In a Topos, this means that the composition of ⋒ and the product-arrows $\langle T, F \rangle$, $\langle F, T \rangle$, $\langle F, F \rangle$ and $\langle T, T \rangle$ is equal to the arrow T only in the last case, and is equal to the arrow F in the other cases.

That ⋒ $\circ \langle T, T \rangle = T$ immediately follows from the very definition of conjunction as centration of $\langle T, T \rangle$:

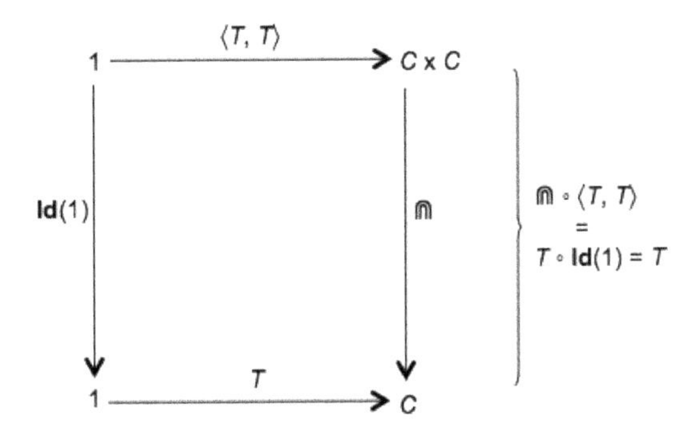

I leave the other cases to the sagacity of the reader. They can be treated in the same style (by the construction of appropriate pullbacks), but they actually relate to more profound considerations on the internal logic of Topos, and in particular on the algebraic structure of sub-objects, which allows us to clarify the action of the Central Object.

It suffices here to show *how* these logical arrows are introduced.

And note that once we have the action of conjunction, there is no reason we cannot make compositions with negation, so that for two arrow elements of C, we have something like ∾ ∘ ⋒ ∘ ⟨T, F⟩. We shall gradually see how all the statements of logical calculus are expressed by the arrows of Topos.

c Exponentiation

Let's show that a^0 is isomorphic to 1 in any Topos.

The diagram of exponentiation gives us:

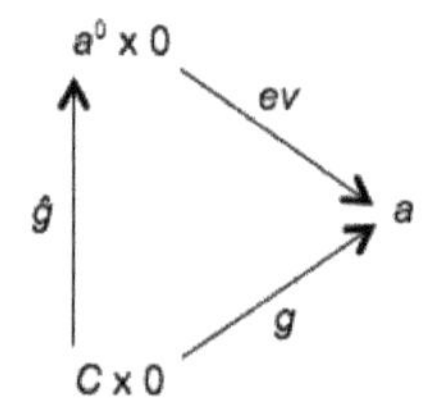

Remember that there are exactly as many arrows $\hat{g}$ between C and a^0 as there are arrows g between $C \times 0$ and a. But we have shown (§9) that $C \times 0$ is isomorphic to 0. Therefore $C \times 0$ is initial, and there is only one arrow g. Thus, for any object C, there is finally only one arrow from C to a^0. The result is that a^0 is terminal, and therefore isomorphic to 1.

Moreover, the exponentiations by 0 give us nothing new: $C^0 = 1^0 = 1$.

What can be said of the exponentiations by 1?

When we recall that for any a, $a \times 1$ is isomorphic to a, the diagram of exponentiation becomes:

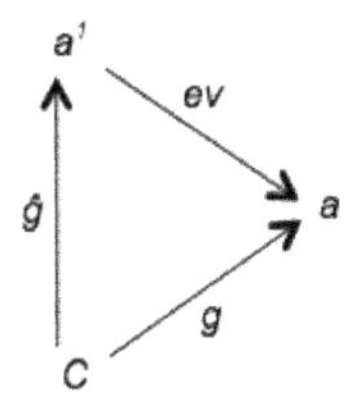

It is thus a useful exercise to show that there is an isomorphism between a^1 and a (using, as arrow g, the identity of a).

At the end of the day C^1 is isomorphic to C, just as 1^1 is to 1. Ever-unproductive operations.

It is once again exponentiation by C that will force us to admit new objects. And yet again, not in every case! It is obvious that 0^c is isomorphic to 0. But what can be said of 1^c? The diagram is:

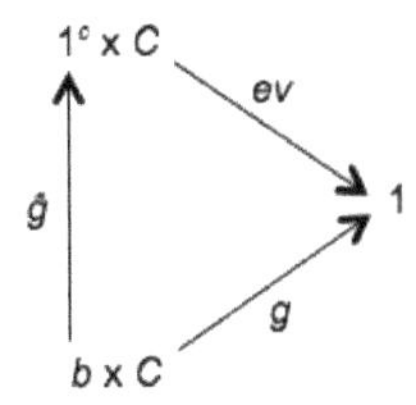

It is clear that, 1 being terminal, there is only one arrow g from $b \times C$ to 1^c. It follows that 1^c is terminal, and therefore isomorphic to 1.

It is then finally C^c which presents itself as a new categorial object.

d Central Object

The centration of our 'mandatory' arrows gives us these old acquaintances.

- Centration of the arrow **Id**(1): this is the true.
- Centration of the arrow $0 \dashrightarrow 1$: this is the false.
- Centration of the true arrow: this is **Id**(C).
- Centration of the false arrow: this is the negation.
- Centration of the arrow **Id**(0): this is the unique arrow $0 \dashrightarrow C$.

It is above all the arrows from C to C which will open up centrations unidentified in the list above. Take for example the identity of C. The diagram is:

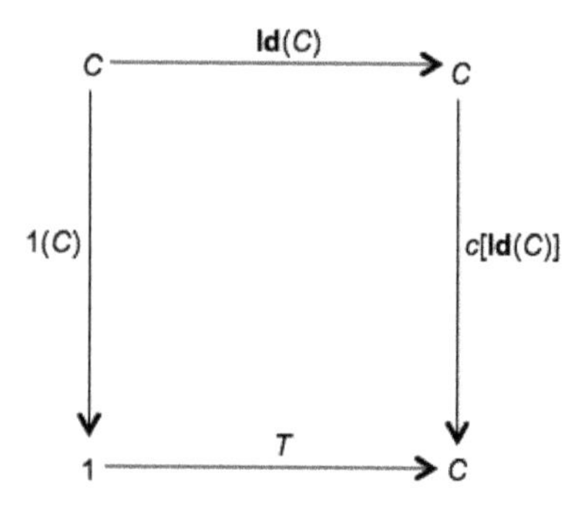

The commutation shows that $c[\mathbf{Id}(C)]$ must be equal to $T \circ 1(C)$. This expression $T \circ 1(C)$ is however interesting, as it is *the composition of the count-for-one of the object and the true arrow.* In fact, the unique arrow $1(C)$ 'projects' C in 1, and recapitulates it as a pure point with no element other than itself (since 1 has one element, namely, the arrow 1···►1, 1 is not empty, but this element is nothing other than $\mathbf{Id}(1)$, meaning that 1 is basically 'filled by itself'). This is what we will call the count-for-one of C. The true test of this count is the composition $T \circ 1(C)$. This is the truth of the object *as the one that it is*. It should be noted $T(C)$, or 'truth of C'. And we will state that the centration of the identity of an object is the truth of that object.

18 ONTOLOGY OF THE VOID AND DIFFERENCE

We remarked above (§13) that categorial formulation, in sharp contrast with set-theoretic exposition, does not require the uniqueness of the void.

This point is of the greatest importance because *the univocity of the name of the void is what gives the set-theoretical discourse on multiple-being its proper unity.*

The hierarchy of pure forms of the multiple originates in the void, and only *names* the schemas of being *qua* being by composing itself with the name of the void (the void set, ∅), which is why I have proposed to alternatively call it 'the proper name of being'.

It is essential to see that the uniqueness of the void *results* from a thought of difference, from an articulation of the principles of thought and of the multiple given at the *point* of difference.

A decision of thought crucial to the development of set theory ontology (which is to my mind still today the only consistent ontology that I know of) is in fact the following: *if two multiples are different, then there is a local determination, or a point, of this difference.* This is the real content of the axiom of extensionality, thus of an axiom of immanence. Because if two sets are only equal if they have the same elements, it follows that two sets are different *if at least one element belongs to one and not the other.* You therefore always make a decision about a global difference (between two sets) on the basis of a local difference (the demonstration of the existence of *one* element presented in the one and not in the other).

This is a radical thesis on the thought of being that poses that any difference between two beings (two multiples) transpires *in a point.* This thesis excludes that two beings only differ globally or qualitatively, or by their general and recapitulative 'strength'. We might say that being's law of donation, or of its thinkable exposition, is that its differentiation is itself always presented locally. Or even: every difference proposes a localization of differentiation.

The uniqueness of the void is immediately inferred from this point. Because the void is that which presents no element, no point-of-itself, and therefore its differentiation with 'another void' is unthinkable, since in this case we would necessarily have a global difference with no local measure, a purely qualitative difference between two 'colors' of the void.

In the end, set-theoretic ontology ties together a thesis of immanence (a multiple is identified by what belongs to it), a thesis on difference (it always presents itself to thought as a point of itself), and a founding thesis on the void as a proper name which is at once originary and sufficient.

The *general* concept of possible universes, as categorial thought delivers it under the name of Topos, doesn't take up any of these three theses. An object is identified (including what concerns its quasi-immanence, its sub-objects) only by its external relations with other objects of the universe (the arrows). A difference (between two objects, between two arrows) may well have no elementary assignation, thus it may be purely global and qualitative (or even be a simple literal difference, as in the case of two isomorphic objects). And finally there can be numerous empty objects (ontologically void, without elements), 'numerous' beings taken here in the categorial sense: numerous non-isomorphic voids.

In this space of possible universes, in these 'logics of universes' that is categorial formulation, the *singularity* of set theory ontology will forcefully reassert itself. The connections of thought this ontology practices 'unthinkingly' (without thinking its proper thought) will be clearly designated as characteristics of *some* possible universes.

This is how category theory allows us, not as a decision as to the thought of being, but as an open exploration of the underlying logics of *every* decision of this kind, to think the thought of being.

On the point at hand, this capacity for reflexive clarification is given in a theorem which essentially says: if a Topos admits – as the universe of sets – the localization of all difference, then the void is unique, and *it is necessarily in the initial position*. Category theory thereby changes an observable characteristic of the set-theoretic universe into a logical law applicable to 'all' possible universes that 'resemble' (with regard to this or that principle of thought) the set-theoretic universe.

The first step in this direction is to categorially define what is meant by 'the elementary mark of all difference'.

We will say that two different parallel arrows f and g are *elementally different* (or different 'at a point') if an element of their source x exists such that: $f \circ x \neq g \circ x$:

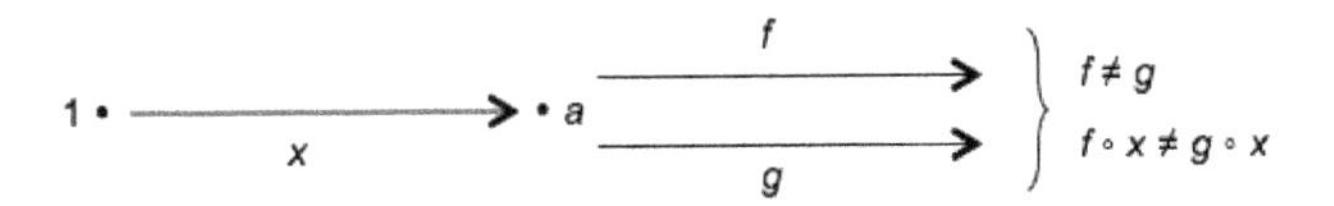

We see that the elementary arrow x (arrow from 1 to a, common source of f and g) serves as local validation for the overall difference between actions f and g. Because the difference of f and g is in some way attested by the difference of their 'applications' to the element x of their source.

Projecting this characteristic onto the scale of an entire possible universe, we will say that a Topos is *well-pointed* if, in this Topos, two different parallel arrows are always elementarily different.

We can thus say that a well-pointed Topos (the name is well chosen by mathematicians, as it 'points' to what is in question: that every difference transpires in a point) is a possible universe which resembles the set-theoretic universe with regard to the thesis on difference. 'Well-pointedness' provides the *logic* of difference, such that it structures the thought of multiple-being, as the singular logic of one kind of possible universe, namely, a Topos which has this property.

What is impressive however is that this logic immediately prescribes (this is the movement of thought typical of Categories) its ontological consequences: *If a Topos is well-pointed, every empty object therein is isomorphic to an initial object 0*. Here we demonstrate the general connection between the thesis on difference and *two* theses on the void: its uniqueness (logically, and thus categorially, this will be 'up to isomorphism') and its foundational (initial) position.

What is especially striking is that this demonstration necessarily depends on the Central Object. This is further evidence that it is *this object* that confers on categorial universes their power of logical clarity.

Take a well-pointed Topos which is not degenerate. Now consider a *non-initial* object a, which means: not isomorphic to zero. Such an object necessarily exists, since the Topos is not degenerate. Similarly necessarily existent (cf. our exercises in §17 on the three objects 0, 1 and C) are the arrows 0_a (from the initial object 0 to a) and **Id**(a) (from a to a).

These two arrows are monomorphisms with the same target a (that 0_a is a monomorphism is the result of there being no arrow prior to 0 except **Id**(0); that **Id**(a) must also be one is trivial, although it must be proven). They are therefore two sub-objects of a. And as there is *no* isomorphism between their sources (since a is presumed non-isomorphic to 0), these sub-objects are really 'two': they are not similar.

But two dis-similar monomorphisms have effectively different (in the strict sense) centrations. All of this is contained in the fundamental theorem of the Centre (cf. §16). Yet these centrations also necessarily exist in a Topos. We have therefore produced two different arrows, $c(0_a)$ and $c[\mathbf{Id}(a)]$. These arrows are obviously parallel, since they go from a to C (the Central Object). Let us recall (to speed up the learning process) the diagrams:

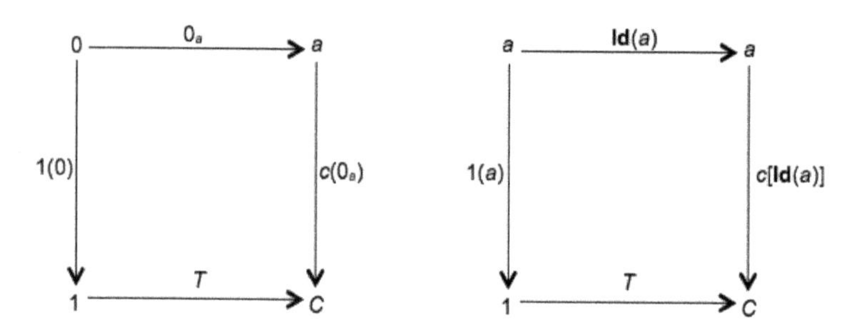

We are therefore in a position of having, in any Topos, two definitely different parallel arrows.

At this point we should remember we have assumed our Topos to be well-pointed. Therefore, the overall difference of these two arrows necessarily affirms itself through the existence of an element x of their common source which confirms this difference, giving us the following situation:

$$1 \xrightarrow{\;x\;} a \underset{c[\mathbf{Id}(a)]}{\overset{c(0_a)}{\rightrightarrows}} C \quad \left.\begin{matrix} c(0_a) \circ x \\ \neq \\ c[\mathbf{Id}(a)] \circ x \end{matrix}\right.$$

But this requires that a has at least one element, the element specified by the name x, which locally differentiates $c(0_a)$ and $c[\mathbf{Id}(a)]$. Consequently, a is not empty.

We have thus demonstrated that *if* a is not isomorphic to zero, *then* it is not void. This can also be said as follows: if it is empty, then it is isomorphic to zero.

In a well-pointed Topos, 'void' and 'zero' (or initial) turn out to be concepts with the same extension, and are therefore interchangeable with each other. And as we know that there is only one zero (in the categorial sense: all the zeros, or all the letters which name them, are isomorphic), we must conclude that there is only a single void.

Thus the foundational (initial) uniqueness of the void is itself the ontological effect of the thesis on the thought of difference. From the logical (thesis on difference) to onto-logical (thesis on the void), the (categorial) result is valid.

19 MONO., EPI., EQU., AND OTHER ARROWS

In the preceding chapters we have seen what the concept of Topos is logically capable of, not least with respect to the fact that it constitutes an illuminating breakthrough in the thinking of the theory of multiple-being.

Our goal however is rather more ambitious. We will eventually come to some crucial onto-logical considerations. But to do this, we must return to the Topos as possible universe and familiarize ourselves with its different kinds of arrows and their different behaviors.

In this chapter we demonstrate eight theorems, which are very important tools for what is to come. This is also the perfect occasion for a recapitulation of some concepts we have already defined.

The first four theorems concern the broader context of closed Cartesian categories, or categories *tout court*.

The following three concern Topoi in general.

The last concerns well-pointed Topoi.

Theorem 1 (all categories): An isomorphism is always equally a monomorphism and an epimorphism.

Let i be an isomorphism, and j its inverse. Consider the diagram:

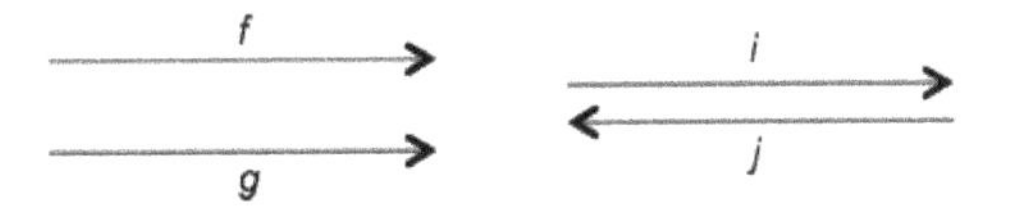

Let's assume that $i \circ f = i \circ g$. We then have:

$$j \circ i \circ f = j \circ i \circ g$$

But since j is the inverse of i, we have

$j \circ i = \mathbf{Id}(b)$, and therefore

$\mathbf{Id}(b) \circ f = \mathbf{Id}(b) \circ g$, thus

$f = g$

If $i \circ f = i \circ g$, this is because $f = g$. The arrow i cannot by itself introduce an identity which did not previously exist. It conserves differences: it is a monomorphism. The demonstration that i is an epimorphism is exactly the same, but 'in reverse'. It is the dual of the preceding demonstration (on duality cf. §6).

Note. We have just shown that an isomorphism is a monomorphism *and* an epimorphism: it conserves and preserves differences. But we have *not* shown the reverse: that every arrow which is at once an epimorphism and a monomorphism is an isomorphism. We are not able to *unify* the categorial principle of the One (isomorphism) and the capacity to preserve *and* conserve differences. In other words, while it is true that identity (isomorphism) implies that differences are conserved and preserved by the indentificatory action, it had not been established that the addition of these conservations of differences (retrospective and anticipatory, depending on whether it is a monomorphism or epimorphism) equals identity.

This point is philosophically important, because it deals with the possibility of *defining* identity from the relation of differences.

In fact, in any given category, it is generally *not true* that every epic and monic arrow (utilizing these charming abbreviations) is isomorphic. Our 5th theorem shows however that it is always true *in a Topos*.

Theorem 2 (all categories): Every equalizer is a monomorphism.

An equalizer (cf. §5) is the limit of a diagram made up of two parallel arrows.

Consider the following diagram, in which e is the equalizer of two arrows f and g, and where we assume that prior to e are two parallel arrows t_1 and t_2 such that $e \circ t_1 = e \circ t_2$ (the goal is to show that $t_1 = t_2$, thus proving that e is a monomorphism):

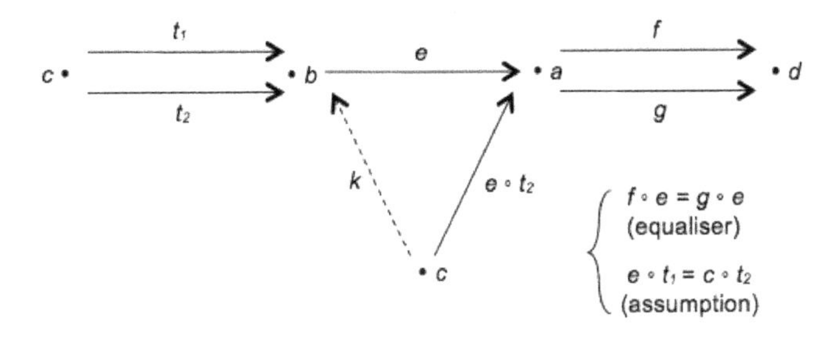

Note. The object c is marked twice in the diagram, to make it clear that $(e \circ t_2)$ equalizes f and g – as we will show – and that therefore, by virtue of the limit rules, there is indeed the unique arrow k from c to e.

We have:

1	$f \circ (e \circ t_2) = (f \circ e) \circ t_2$	(associativity)
2	$(f \circ e) = (g \circ e)$	(e is the equalizer)
3	$f \circ (e \circ t_2) = (g \circ e) \circ t_2$	(by 1. and 2.)
4	$f \circ (e \circ t_2) = g \circ (e \circ t_2)$	(associativity)

We see that $(e \circ t_2)$ equals f and g. There therefore exists, since e is an equalizer (thus a limit for the equalizing act), a unique arrow k of c to b which commutes the triangle b–a–c, thus a unique arrow k such that $e \circ k = e \circ t_2$.

In truth the arrow k is t_2 itself, for it is obvious that $e \circ t_2 = e \circ t_2$! And as only the arrow k can verify the equation $e \circ k = e \circ t_2$, it is impossible that k $\neq t_2$.

That being said, we have assumed that $e \circ t_1 = e \circ t_2$. Consequently, t_1 *also* verifies the equation! The uniqueness of its solution therefore requires that $t_1 = t_2$.

Finally, if e is an equalizer, for any pair of parallel arrows t_2 and t_1 such that $e \circ t_1 = e \circ t_2$, we necessarily have $t_1 = t_2$, which proves that e is a monomorphism.

Theorem 3 (all categories): every equalizer which is an epimorphism is an isomorphism.

Take the equalizer e of the parallel arrows f and g. We have $f \circ e = g \circ e$. But then, if e is epic we necessarily have $f = g$ (because e preserves differences, and is therefore 'right-cancellable'). Consider the following diagram:

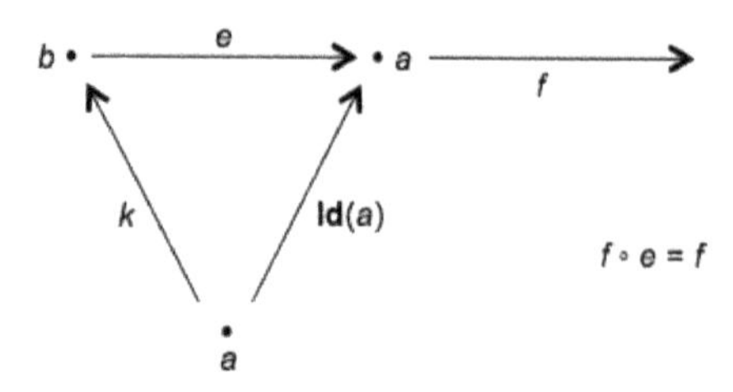

This diagram is legitimate because $f \circ \mathbf{Id}(a) = f : \mathbf{Id}(a)$ equalizes f with itself, as must e. There is thus a unique arrow k from a to b which commutes the triangle.

It then follows:

1	$e \circ k = \mathbf{Id}(a)$	(commutation of the triangle)
2	$(e \circ k) \circ e = \mathbf{Id}(a) \circ e = e$	(by 1.)
3	$e \circ (k \circ e) = e$	(by 2. and associativity)
4	$e = e \circ \mathbf{Id}(b)$	(by identity)
5	$e \circ (k \circ e) = e \circ \mathbf{Id}(b)$	(by 3. and 4.)

But we have shown (theorem 2) that the equalizer e is a monomorphism. So if we have $e \circ (k \circ e) = e \circ \mathbf{Id}(b)$, it is because we already have:

6 $k \circ e = \mathbf{Id}(b)$.

The comparison of equations 1 and 6 shows that the arrow k is the inverse of the arrow e. Consequently, e is an isomorphism. It follows that every equalizer that is an epimorphism, is an isomorphism.

> *Note.* We said just now that while it is consistently true that an isomorphism is monic and epic (that is, iso ···➤ monic + epic), it is at the same time not true in general (in any given category) that an arrow that is both monic and epic arrow is by this very fact iso (that we have monic + epic ···➤ iso). And so we cannot in general pose the equation: monic + epic = iso.
>
> What we have just shown is that if an arrow *that is an equalizer for two parallel arrows* is at the same time monic (as it *always* is, theorem 2) *and* epic, then it is iso.
>
> It follows that if, in a category, every monic arrow is the equalizer of some pair of parallel arrows, it will be true that moni + epic = iso.

Because an arrow that is both monic and epic will be, being monic, an equalizer, and being epic, an epic equalizer, thus an isomorphism.

If for a particular category we wish to show that iso = mono + epic, we simply need to establish that in this category every monic arrow is the equalizer of at least one pair of parallel arrows (therefore the limit of at least of one diagram made up of two parallel arrows).

The essence of this point is that the act of equalization 'at the limit' mediates between identity (isomorphism) and the 'neutral' treatment of differences.

The demonstrations of theorems 2 and 3 clearly show why. On the one hand, an equalizer always conserves differences, *otherwise it would not have the power to anticipate their nullification*, as it does in equalizing two different arrows which follow it, doing so moreover from the universal position. It follows from this that it is a monomorphism. But if it is also an epimorphism, it equalizes not *two* arrows (thus a real difference), but *one arrow with itself*, which brings identity onto the scene. We will also say in this case that it preserves differences, because it does not deal with any. It is this double capacity to preserve every difference and to do so only under the sign of identity which finally identifies it as an isomorphism.

Theorem 4 (all categories): The pullback of a monomorphism is a monomorphism.

Take a pullback whose base arrow (the bottom of the square) is a monomorphism m_1. We will say that the arrow at the top of the square, m_2, is the monomorphism's pullback. The goal is to establish that this arrow is also a monomorphism.

Consider the following diagram:

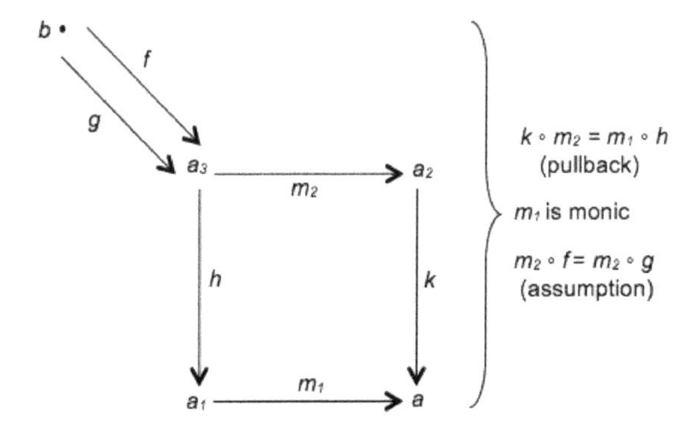

Our assumption is that $m_2 \circ f = m_2 \circ g$, and our aim is then to show that when $f = g$ this will establish that m_2 is a monomorphism.
It follows that:

1	$k \circ m_2 = m_1 \circ h$	(commutation of the square)
2	$(k \circ m_2) \circ g = (m_1 \circ h) \circ g$	(by 1.)
3	$k \circ (m_2 \circ g) = m_1 \circ (h \circ g)$	(associativity)
4	$k \circ (m_2 \circ f) = m_1 \circ (h \circ g)$	(hypothesis regarding *f* and *g*)

We see that the arrows $(m_2 \circ f)$ and $(h \circ g)$ commute an 'exterior' square, as the following diagram shows:

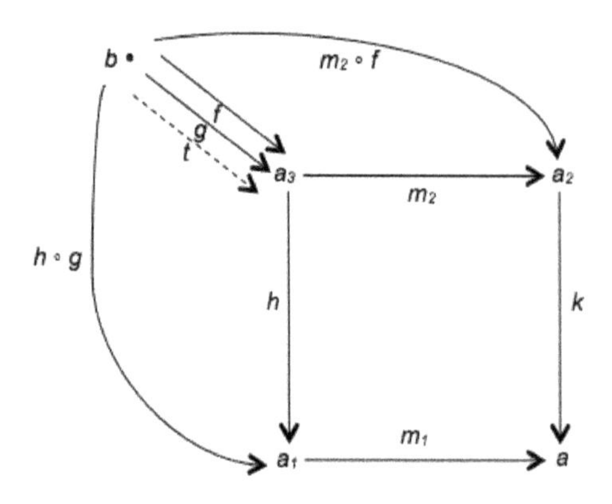

Since the interior square is a pullback, there is therefore a *unique* arrow *t* from *b* to a_3 that commutes the entire diagram.
Specifically, we have:

$$m_1 \circ (h \circ t) = m_1 \circ (h \circ g)$$

But since m_1 is a monomorphism it is left-cancellable, and we have $h \circ t = h \circ g$. Therefore the arrow *g* (like *t*) also commutes the triangle b–a_3–a_1. Only it turns out that *g*, again, like *t*, also commutes the top triangle b–a_3–a_2, since $m_2 \circ g = m_2 \circ f = m_2 \circ t$. The arrow *g* commutes the entire diagram, by which it follows, by reason of the uniqueness of the arrow with this capacity, that $t = g$.

An exactly identical calculation, replacing $m_2 \circ f$ at the top of the diagram with $m_2 \circ g$ which we presume to be equal would show that we equally have (these are merely nominal substitutions) $t = f$.

Finally, we have $f = g$, from which we suppose (m_1 being a monomorphism) that $m_2 \circ f = m_2 \circ g$. It follows that m_2 is a monomorphism, which demonstrates theorem 4.

Theorem 5 (Topos): In a Topos, every monomorphism m of target b (therefore every sub-object of b) is the equalizer of its centration $c(m)$ and of the arrow $T \circ 1(b)$, where T is the truth arrow, and $1(b)$ is, as always, the (unique) arrow from b to 1.

Note. Observe that:

- On the one hand, that we follow the strategy announced in theorem 3 for establishing that in categories of a certain type (here, Topoi) we indeed have: iso = monic + epic. The next step is to show that every conserver of difference is an equalizer, and thus to find the pair of arrows that it equalizes in the limit position.
- On the other hand, that the arrow $T \circ 1(b)$ is the special arrow (cf. §17.) that we have agreed to note as $T(b)$, or 'truth of b', that which combines the count-for-One of the object b and the true arrow. Thus we can formulate the theorem philosophically as follows: *every arrow that conserves differences is the equalizer of its centration and of the truth of its target.*

We begin with a small and trivial lemma: given an arrow f from a to b, we necessarily have $1(b) \circ f = 1(a)$. In other words, the triangle below commutes:

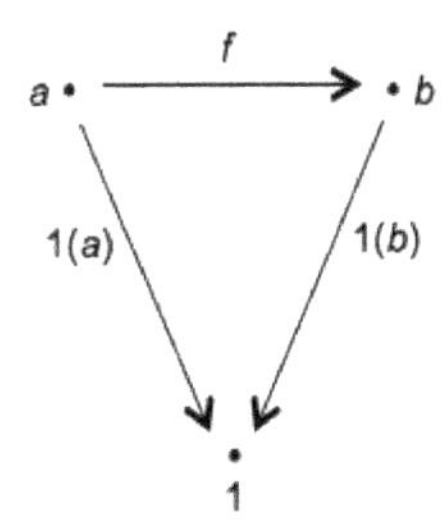

The point is that $1(b) \circ f$ is an arrow from a to 1, and there is only one arrow going toward a terminal object, which is $1(a)$.

Now, consider the centration pullback of m, and the arrow $1(b)$ it bears:

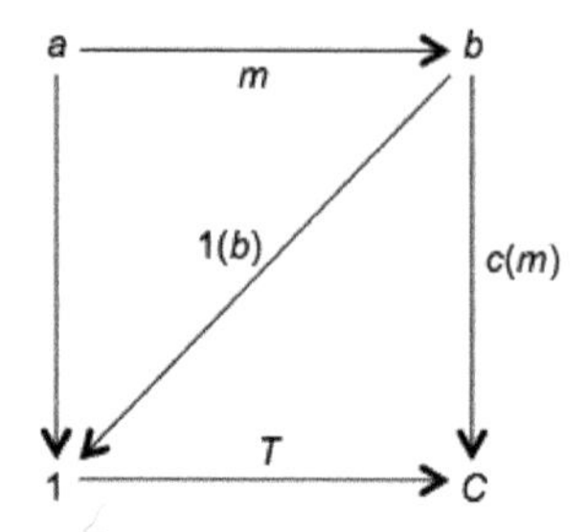

It follows:

1 $c(m) \circ m = T \circ 1(a)$ (commutation of the square)
2 $1(a) = 1(b) \circ m$ (lemma above)
3 $c(m) \circ m = [T \circ 1(b)] \circ m$ (by 1. and 2.)

Equation 3 indicates that the arrow m equalizes the arrows $c(m)$ and $[T \circ 1(b)]$. It remains to show that this equalization is a limit. This requires us to consider what happens if an arrow other than m, say f from source c and target b, also equalizes $c(m)$ and $[T \circ 1(b)]$. As in the following situation:

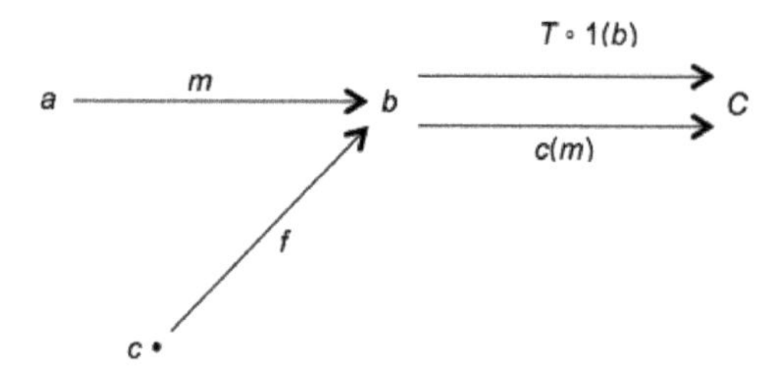

It then follows:

1 $T \circ 1(b) \circ f = c(m) \circ f$ (assumption)
2 $1(b) \circ f = 1(c)$ (lemma)
3 $T \circ 1(c) = c(m) \circ f$ (by 1. and 2.)

Equation 3 indicates that the arrows leaving from c in the following diagram commute the exterior of the square:

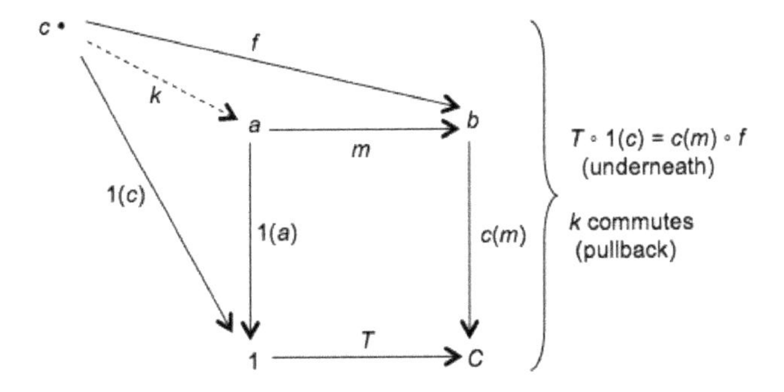

But the interior square is a pullback. There must therefore be a unique arrow k from c to a which commutes the entire diagram, and specifically the triangle c–a–b. Which, reduced to the diagram of equalization, gives us:

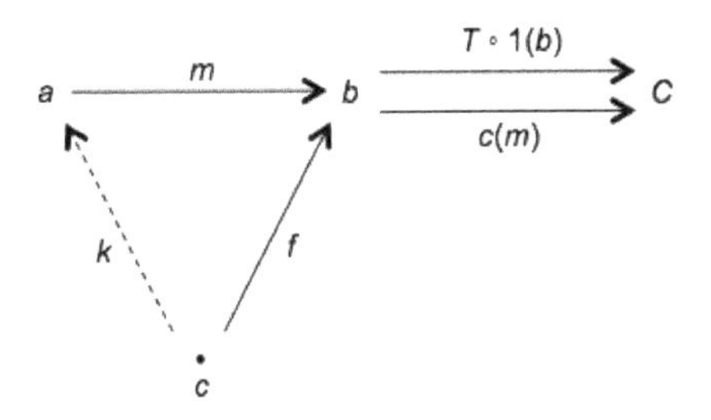

Which clearly shows that the arrow m is in the limit position for the equalization of $c(m)$ and $T \circ 1(b)$, this is to say of $c(m)$ and of $T(b)$. Therefore it is verified that in a Topos, every monic arrow is the equalizer of its centration and of the truth of its target.

Theorem 6 (Topos): In a Topos, iso = monic + epic.

We know (theorem 1) that every iso arrow is monic + epic.

We know (theorem 3) that every epic equalizer is iso.

But theorem 5 tells us that, in a Topos, every monic arrow is an equalizer. Consequently, every monic *and* epic arrow is an epic equalizer, therefore, by theorem 3, an iso arrow.

In a Topos, every iso arrow is monic and epic (this is true in any category), and reciprocally every monic and epic arrow is iso. Consequently, in a Topos, iso = monic + epic.

This is undoubtedly a very profound characteristic of Topoi, this knotting (as already noted) of the treatment of differences and identity. Isomorphism, categorial form of the One, can be analysed in terms of the conservation and preservation of differences. We will say that in a topological universe, the One has an *analytic* status.

Theorem 7 (Topos): In a Topos, every arrow admits a decomposition into an epic arrow followed by a monic arrow. Or alternatively, given f, there exists an m (monomorphism) and an e (epimorphism) such that $f = m \circ e$.

The monomorphism of this canonical decomposition of f is called the *image* of f, and is written im(f).

We cannot overestimate the importance of this theorem, because it shows that the analysis of iso arrows in terms of monic and epic arrows extends, in a certain sense, to any arrow whatsoever. We can then say that an arrow is always equivalent to a preservation of differences (epic) followed by a sub-object of its target (monic). The situation is as follows:

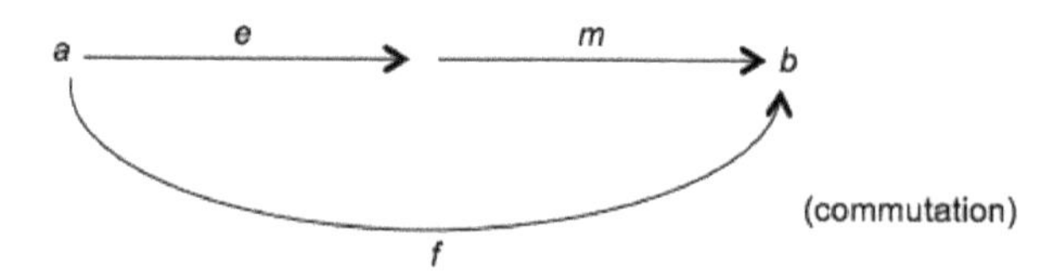

Here as well, the 'active' identity that represents an arrow can be analysed in terms of the preservation and conservation of differences.

We have seen for some time now to what extent identity, in the categorial formulation, is evasive, fleeting, 'up to isomorphism', etc. The essence of the question is that *identity is analytically subordinate to difference*, which theorem 7 is again an illustration of.

But the price paid for this subordination is the admission of global differences which have no elementary testing ground: most Topoi *are not well-pointed*. Theorem 8 will give us a first insight into the 'exceptional' character of well-pointed Topoi.

To obtain the monomorphism 'latent' in any action f, we still have to make use of an equalizer. The construction is convoluted, but conclusive.

Consider the *pushout* of the arrow f with itself. The pushout is the limit of the diagram of two arrows that have the same source (the pullback is the limit of the diagram of two arrows which have the same target). This pushout is presented as such:

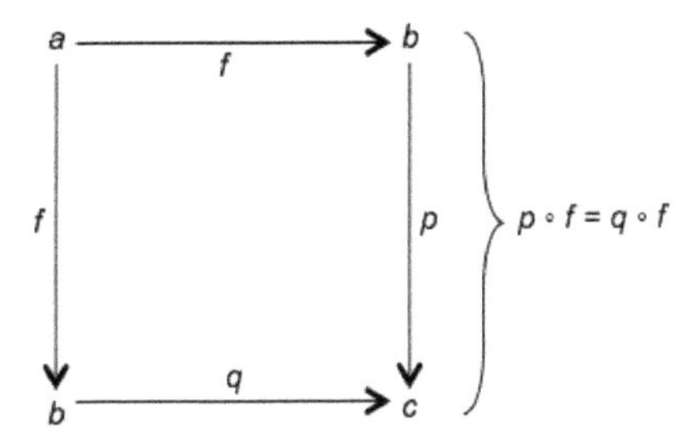

Now consider the equalizer im(f) of arrows p and q (which are parallel arrows, each going from source b to target c). This is an arrow with target b, and, say, with source d. We evidently have $p \circ \text{im}(f) = q \circ \text{im}(f)$. But the commutation of above pushout also gives $p \circ f = q \circ f$, so that f also equalizes p and q. Thus (universal position of the equalizer), a unique arrow e exists from a to d which commutes the triangle d–a–b:

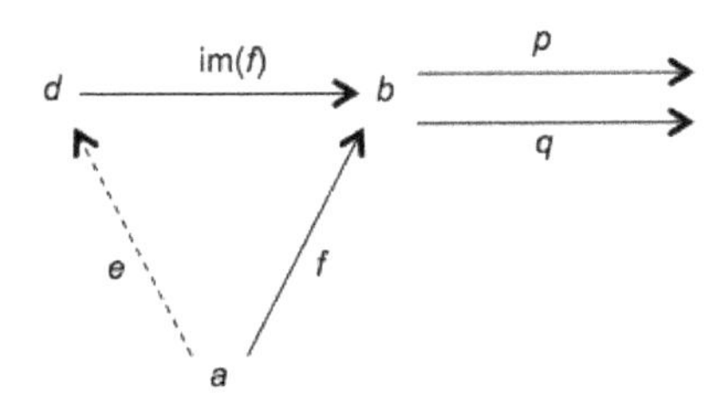

Thus, we necessarily get:

$f = \text{im}(f) \circ e.$

Yet, we know (theorem 2) that every equalizer is a monomorphism. Therefore, im(f) is a monomorphism, and the first part of our work is completed.

We still have not shown that e is epic. This is tedious work, which I leave for another time. It is over the course of this labour that the characteristics of Topos come into the picture (via theorem 6 in particular, which affirms that in a Topos, every monic arrow is an equalizer), since the above demonstration, which shows the monic component of the analysis of f, applies to all closed Cartesian categories.

This labour yields what we expect: e is epic, and moreover, every arrow in a Topos can be analysed in the composition of an epimorphism and a monomorphism. We will call this analysis (which the *current demonstration* shows is unique up to isomorphism) *the epi-monic factorization of f*.

Theorem 8 (well-pointed Topos): All well-pointed Topoi are bivalent. This means that its Central Object only has two elements: the true and the false.

This fascinating theorem is a veritable interweaving of logic and ontology.

Particularly in sections 15, 16 and 17 we lifted a corner of the veil on the linguistic and logical functions of the Central Object of a Topos. We have begun to show that it was not for nothing that the element C which serves as the basis for centration pullbacks has the true for its name; that in direct opposition to this 'verifying' this action is the action of the false; that between the two there is the action of the arrow of negation, and that we have for example $\infty \circ T = F$. We have thus brought to light the singular role of the elements of the Central Object (the arrows from 1 to C) which appear to us more and more as *evaluative actions*, on the basis of the centration of monic arrows (or sub-objects). In particular, we have seen that the 'membership' of an element of b to a sub-object f_p intended to 'mark' in an object b a property P (given as centration) can be ascertained from the equation $P \circ x = T$.

A bivalent Topos expresses a logic with some features at least one of which we are familiar with: actually, there are but two evaluative actions, the true and the false. It is a well-known fact this contrasts with modal logic, which admits, beyond the true and the false, other evaluative norms and statements such as the possible, the necessary, the probable, etc., giving us logics that are tri-valent, multi-valent, and finally, extremely nuanced logics which possess an infinity of norms.

Still, classically, and especially in set theory ontology, there are only two values for any statement.

We will note that in a non-degenerate Topos the true and the false necessarily exist, as we have seen, and are different (dis-similar) arrows. In this sense, bivalence is a 'minimal' property: it limits the number of possible values to that which the general structure of a Topos renders necessary: two values, the true and the false.

Yet what we have shown is that bivalence is *constrained* by the ontology of difference: *if* every difference admits a localization (well-pointed Topos), *then* the Topos is bivalent, C has only two elements.

The demonstration is of great interest. Take a well-pointed Topos, and consider an element f of the Central Object C (thus an arrow f from 1 to C). The pullback of f and of T necessarily exists, because these two arrows have C as their target. Thus we have the following square, which is a pullback:

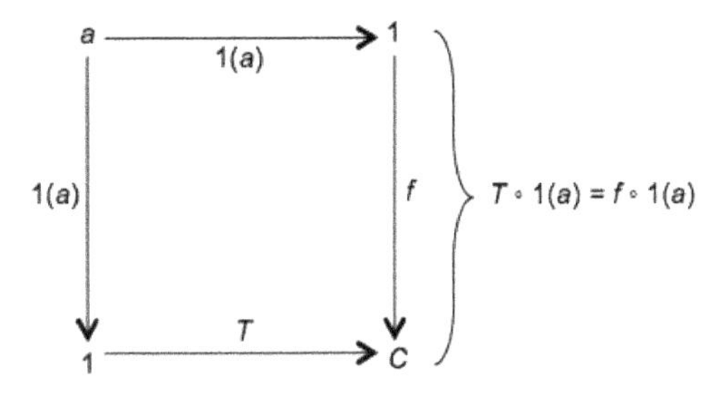

The arrow $1(a)$ at the top of the square is the pullback of a monomorphism. In fact, the bottom arrow, T, going from 1, is a monomorphism (like every elementary arrow). However (theorem 4), the pullback of a monomorphism is a monomorphism. This means that the arrow $1(a)$ is a monomorphism, thus a sub-object, and that *the pullback is a centration pullback*. The hypothetical element f of C is in fact the centration of $1(a)$.

We thereby distinguish two cases.

- **a** Object a is isomorphic to 0. In this case, the arrow $1(a)$ is similar to the arrows $1(0)$. These two arrows therefore have the same centration (fundamental theorem of the Central Object). Yet, by definition it is the same for the false; the centration of $1(0)$ is the false, and we have just seen that the centration of $1(a)$ is f. Thus $f = F$.
- **b** Object a is not isomorphic to zero. In this case, since the Topos is well-pointed, a is not empty. This is the central tenet of section 18: in a well-pointed Topos, the elementary localization of every difference requires that every empty object *must* be isomorphic to 0.

But if a is not empty, possessing an element x, $1(a)$ is always epic. Let's consider the diagram:

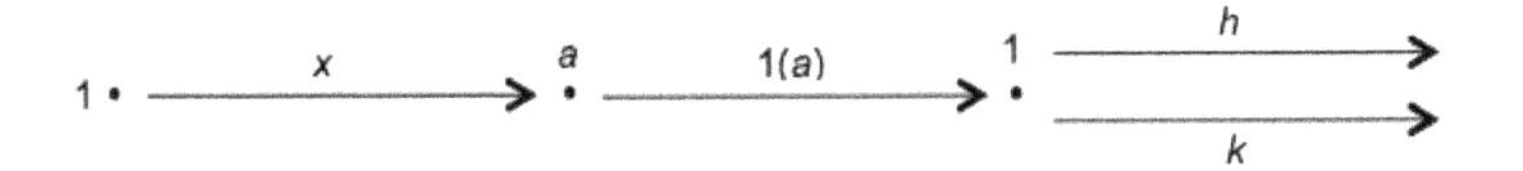

Let's assume that $h \circ 1(a) = k \circ 1(a)$. It then follows:

$$h \circ 1(a) \circ x = k \circ 1(a) \circ x$$

But $1(a) \circ x$ is an arrow from 1 to 1, therefore none other than $\mathbf{Id}(1)$, since 1 is terminal (as noted several times already). Therefore:

$$h \circ \mathbf{Id}(1) = k \circ \mathbf{Id}(1)$$

from which it obviously follows that $h = k$.

By showing that if $h \circ 1(a) = k \circ 1(a)$, then $h = k$, we have shown that $1(a)$ is epic.

Finally, $1(a)$ is both epic and monic. And as we are in a Topos, theorem 6 tells us that $1(a)$ is an isomorphism between a and 1. It follows that the arrow $1(a)$ is similar to the arrow $1(1)$, that is to say $\mathbf{Id}(1)$. But two similar arrows have the same centration. Now we know that the centration of $\mathbf{Id}(1)$ is nothing other than T, while that of $1(a)$ is f. Thus $f = T$.

Ultimately, it turns out that our hypothetical element f of C can only be either F (if a is isomorphic to 0), or T (if a is not isomorphic to 0). The assumption of a third evaluative action, as an element of C, fails. The Topos is bivalent.

This time, it is ontology that prescribes logic, and all the more so since the heart of the proof mobilizes the central ontological property of well-pointed Topoi: the uniqueness of the void. The theorem can in fact be formulated as follows: in a Topos, if the void is unique, the logic is bivalent.

This ontico-logical connection between void and bivalence is fundamental. It indicates that in supposing, or deciding, that being presents itself to thought as the proper name of the void, you compel yourself to recognize that when it comes to evaluative actions, there is only the true and the false.

Note. We have shown that every well-pointed Topos is bivalent. We have *not* shown that every bivalent Topos is well-pointed! This is in fact false. There may be bivalent Topoi that are in no way well-pointed. In other words: the ontological trait (localization of difference, uniqueness of the void) prescribes a logical trait (bivalence). But this logical trait does not prescribe the ontological trait.

The exact position of bivalence with regard to the logic of the universe and its knotting to ontology will only be clarified when we examine other logical characteristics of Topoi, especially their 'Boolean' and 'intuitionistic' classifications. This will be the challenge of the next several sections.

20 TOPOI AS LOGICAL PLACES

That the Central Object prescribes a Topos's logic will be a general thesis about logic, on the following basis:

a In any Topos it is possible to define the value of a statement in any formal language whatsoever, by assigning to it an element of the Central Object.

In other words, for every statement p you associate a possible value of p by making it correspond to an arrow $1 \xrightarrow{f} C$ in the Topos.

We call this procedure an *evaluation*. This is a function V operating between the letters representing statements and the elements of C, which associates an element of C to every letter p that represents a statement, as follows: $V(p) = 1 \xrightarrow{f} C$.

If the Topos is bivalent (if it only admits the true and the false as elements of C) we return to the standard semantic procedures for calculating these propositions: to any letter of the statement we can assign either the value true or the value false.

This will always be the case if the Topos is well-pointed (since we have shown in §19 that every well-pointed Topos is bivalent).

If the Topos admits elements of C other than the true and the false, the interpretation will be of a different kind, containing for example modalities (the possible, the necessary…), or, if C has a very large number of elements, evaluative nuances which have no name in any language.

Nevertheless we will see that the true and the false always have a singular, and therefore structural, position amongst the elements of the Central Object.

b In any Topos it is possible to define logical arrows that interpret (on the basis of an evaluation *V*) the conventional logical connectives, namely negation (not-*p*), conjunction (*p* and *q*), disjunction (*p* or *q*) and implication (*p* implies *q*).

We have already shown in §15 how this works with negation. In every Topos there is an arrow $\backsim$, negation, defined as the centration of the false: $\backsim = c(F)$. We have also shown that the composition of this arrow with the true and the false gives the predictable result, namely, $\backsim \circ T = F$ and $\backsim \circ F = T$.

If for example an evaluation *V* assigns to a statement *p* the value *F* (the arrow *F* from 1 to *C*), the evaluation of the negation of *p* will be the arrow $\backsim \circ F$, which is to say the true, as required by the conventional semantics of negation.

We have also shown in §17 how to define the conjunction arrow $\Cap$, which, in any Topos, will interpret the logical connection 'conjunction' (this arrow is defined as the centration of the product arrow $<T, T>$). So for an evaluation *V* which assigns to both *p* and *q* the arrow *T*, the evaluation of the conjunction '*p* and *q*' will be made by the categorial composition $\Cap \circ <T, T>$, namely, the composition of the conjunctive arrow and the product arrow of the arrows which evaluate *p* and *q*. We have shown that $\Cap \circ <T, T> = T$, which is consistent with the fact that, in conventional semantics, '*p* and *q*' is true when *p* and *q* are themselves both true.

We can also define in any Topos, in a slightly more sophisticated way, an arrow $\Cup$, which interprets disjunction, and an arrow $\Rightarrow$, which interprets implication. But we won't go into this now.

Ultimately, the entire propositional calculus is 'transferred' into the Topos, so that every complex statement, being given a simple evaluation that assigns to the letters *p*, *q*, ... an element of *C* which in turn receives a value, and is 'worth' an element of *C*.

How is the evaluation of complex statements presented? As a series of operations (therefore of arrows like $\backsim$, $\Cap$, etc.), operating *on the elements of C*.

We can then say that logic, semantically evaluated in a Topos, appears *as the unfolding of an algebraic structure on elementary arrows* $1 \rightarrow C$.

Of course, from an immanent point of view, this algebraic structure must be defined entirely in terms of arrows. We have seen for example that what 'connects' the element of *C*

> which evaluates p – namely, $V(p)$ – to the element of C which evaluates q – namely, $V(q)$ – when it comes for example to the evaluation of 'p and q', is the composite arrow $\Cap \circ \langle V(p), V(q) \rangle$, which engages categorial operations like the product arrow and centration. But at the end of the day we can see that through these operations we assign to the couple $V(p)$ and $V(q)$ an element of C (which is precisely the arrow $\Cap \circ \langle V(p), V(q) \rangle$), which is $V(p$ and $q)$, the evaluation of the complex statement. It is thus (seen 'from outside') an operation that assigns to two elements of C a third element of C, thus an algebraic structure.
>
> This is however not immanent to the Topos for a number of reasons, foremost being that the idea of an operation performed on the *elements* of C has no categorial sense, inasmuch as 'the elements of C' (which are a *set* of arrows) are not something accessible from the interior of the Topos. In fact, 'the elements of C' constitute neither an object nor an arrow. While it is a collection of arrows that a hypothetical inhabitant of the Topos would of course see as all going from 1 to C (this is what authorizes the name 'elements of C'), he has no reason to consider that they designate, taken together ('collectively'), an existing entity. Because for this inhabitant, what exists, all that *is*, are objects and arrows.

While the *name* 'element of C' is comprehensible for an inhabitant of a Topos (it designates a kind of arrow), it does not refer to the totality of its real referents. It exists as a concept, and can be used as an identifier on a case by case basis. The universe prescribed by a Topos *is not extensional*: the referent of a name is not the extension of the concept that it names, or, as Frege said, the totality of 'cases' of the concept.

This is a decisive difference – a difference *in thinking* – between the set-theoretical universe and the topological universe. 'The elements of C' is not an existential statement for the inhabitant of the Topos. What exists is *this* arrow, and we can say that it is an element of C. The concept identifies without collectivizing.

The path that we are going to follow is ambiguous. We will speak of the 'structure' of the elements of C, of the operations on these elements, and so on. This is a version that we could say is 'external' to the immanent logic of a Topos. In other words, we will speak of Topos in the language of sets.

It is possible to draw closer still to immanence, and continue to simply (and invariably) trace commutative diagrams (the commutation of a diagram typically being what an inhabitant of the Topos *sees*). But the elucidatory power (external and metaphorical) of set-theoretical language better allows us to better sustain the onto-logical investigation. We must always remember however that many of our expressions, since they refer to the reality of the Topos from the exterior, will be unintelligible for an inhabitant of this Topos.

That is to say that we are adopting the position of what, in *Being and Event*, I called 'the ontologist'.

That said, let's return to the idea that logic is evaluated in a Topos by operations on the elements of C (which are the constituents of the elementary evaluation, that of simple statements or letters).

How then to explore and define, in a properly categorial way (by objects and arrows), a structure on the elements of C, or even, more generally, on the sub-objects of any object of the Topos?

Take as a fundamental example an order-structure, namely, a relation $\leq$ which is reflexive ($x \leq x$) and transitive ($x \leq y$ and $y \leq z \rightarrow x \leq z$).

Is it possible to define an order-structure upon the sub-objects of a given object b?

Take two sub-objects f and g of b. These are both monomorphisms of target b. Suppose that an arrow h exists such that the following diagram commutes:

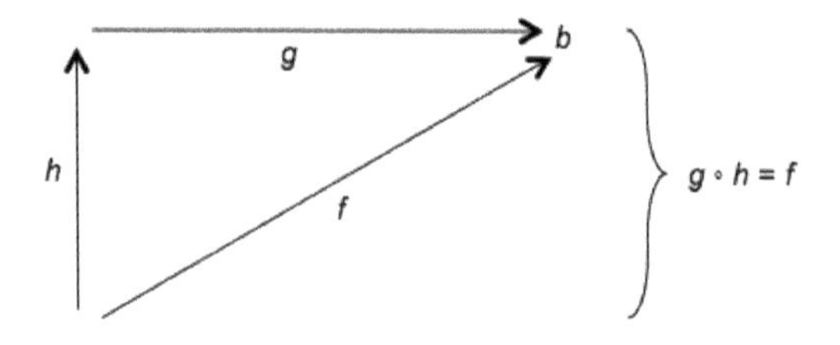

We have $g \circ h = f$. We will then pose that $f \subseteq g$. More precisely: we will say that $f \subseteq g$, or that f is 'inferior to g', if there exists an arrow h such that $g \circ h = f$.

We have already seen that the relation $\subseteq$ is an order-relation.

That it is reflexive is obvious, as $f \circ \mathbf{Id}(a) = f$. We therefore have $f \subseteq f$ ($\mathbf{Id}(a)$ is the h which 'factorizes' f).

That it is transitive can be read from the following diagram, which shows that $f \subseteq g$, and that $g \subseteq h$:

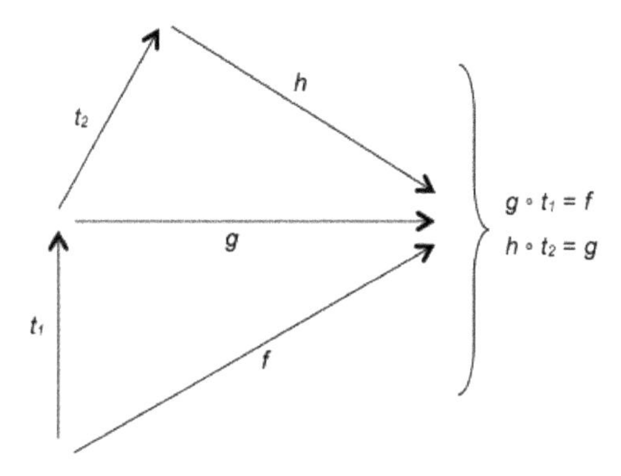

It then follows that:

1	$g \circ t_1 = f$	$(f \subseteq g)$
2	$h \circ t_2 = g$	$(g \subseteq h)$
3	$(h \circ t_2) \circ t_1 = f$	(by 1. and 2.)
4	$h \circ (t_2 \circ t_1) = f$	(associativity)

Equation 4. shows that $t_2 \circ t_1$ factorizes f by h, and therefore that $f \subseteq h$.

Moreover, if $f \subseteq g$ and $g \subseteq h$, we have $f \subseteq h$: the relation is transitive.

So here we have an order-relation between sub-objects of a given object (between monomorphisms with the same target).

Note. This is a *partial* order-structure.

In order that $f \subseteq g$, there must be an arrow h such that $g \circ h = f$. For us to have $g \subseteq f$, there must be an arrow k such that $f \circ k = g$.

Two questions come to mind here:

1 What if *both* cases are present? Take an arrow h which factorizes f by g, and an arrow k which factorizes g by f. It is an excellent exercise that shows that k is the inverse of h, and h is an isomorphism which commutes the diagram. Which brings us back to our old friend: the *similarity* between the arrows f and g (Cf. §11). In other words, f and g are 'the same' sub-object, which, I repeat, is written $f \simeq g$.

We have in fact seen that we have the equation:

$$f \simeq g \leftrightarrow f \subseteq g \text{ and } g \subseteq f$$

2 What if we *have neither case*? There is neither an arrow which factorizes f by g, nor one which factorizes g by f. Well, in that case, f and g *are not comparable*. The relation does not function between these two arrows. And as this case is perfectly possible, we will say that the order we have just defined on the sub-objects of an object is more often than not a partial order.

We are thus in possession of a partial order-structure on the 'set' of sub-objects of an object.

Note. What is 'seen' by the inhabitant of a Topos? He sees that the arrow h is such that $g \circ h = f$. For he knows the network of actions of his universe. He makes no objection to the recognition of the existence of h as criterion for the relation $f \subseteq g$. He can comprehend in what sense this relation is 'ordered'. What he cannot understand is the idea of the existence of a *practical domain* for this relation, because for him the set of sub-objects of an object simply does not exist.

The relation $\subseteq$ is nevertheless a little unwieldy for what remains our essential objective: to investigate possible connections between elements of the Central Object, as the basis for a logical evaluation. Why? Because these different elements *are always in-comparable,* and in no way bound by this relation. This is a variant of the remark we made in §12, where we noted that two similar elements are necessarily identical. To resume, consider the diagram:

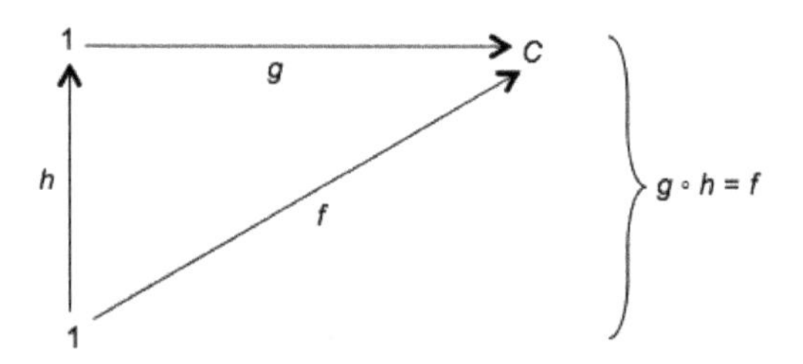

If there is a h such that $g \circ h = f$, the arrow must go from 1 to 1. But the only such arrow is $\mathbf{Id}(1)$. So we have $g \circ \mathbf{Id}(1) = f$, thus $g = f$. The only case where the relation holds is the case of $f \subseteq f$. As soon as $f \neq g$, the relation no longer operates, f and g are in-comparable.

We will overcome this obstacle by drawing on centration. We know there is a bi-univocal correspondence between the 'real' (or dis-similar) sub-objects of b and the centrations of these sub-objects. Consider in particular the sub-objects of 1 (thus monomorphisms of the type $1(a)$). The pullback of the centration of such a monomorphism is:

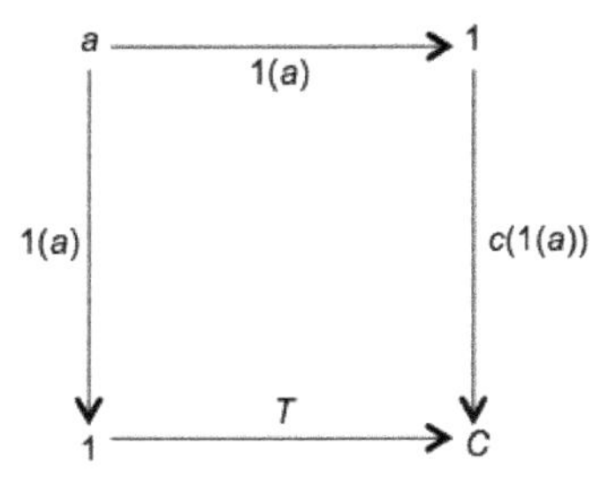

It appears here that *for every sub-object of 1, there corresponds an arrow from 1 to C, which is its centration.*

Inversely, if we take the pullback of the true and an element of *C*, it will appear above a sub-object of 1 whose centration will be the same element of the *C*:

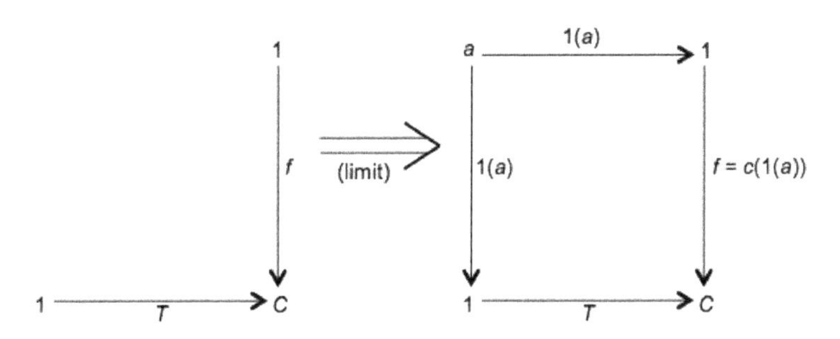

Finally, the elements of *C* and the sub-objects of 1 correspond bi-univocally.

Let us now define a function *J* from the set of sub-objects of 1 on the set of elements of *C*, which 'registers' this bi-univocal correspondence, by making correspond a sub-object $1(a)$ of 1 to the centration of this sub-object, which is an element of *C*:

$$J(1(a)) = c(1(a))$$

Now we can easily (thanks to *J*) 'transfer' the order-relation $\subseteq$ that exists between the sub-objects of 1 onto the elements of *C*. We will effectively define on the elements of *C* a relation $\leq$, which is 'isomorphic' to the relation $\subseteq$, by the equation (f_1 and f_2 being elements of *C*):

$$1(a) \subseteq 1(b) \leftrightarrow J(1(a)) \leq J(1(b))$$

That the relation $\leq$ will have all the properties of the relation $\subseteq$ is guaranteed. Of course, *it won't be defined within category theory in the same way*. It is not the relation $\subseteq$. But all its 'abstract' properties, particularly those expressible by the commutations of arrows, will transfer identically to the relation $\leq$.

Note. We are drowning in set-theoretical vocabulary and confusion is mounting! What is this 'function' J, which is clearly not an arrow of a Topos? And how, since J does not exist *in* the Topos, can we determine that the existence of the relation $\leq$ is really definable through arrows and objects? I fear that in the eyes of the inhabitant of a Topos, our relation between elements of C in no way exists.

It is true. And on this point I have taken the precaution of indicating I was working 'from outside', treating the Topos not as a Universe but as a configuration in *another* universe (in fact, the set-theoretical universe).

This 'universal' ambiguity can however be rearranged and overcome. We can fashion our metaphors from the interior of the Topos to the point where we reach a 'true' order-relation between elements of C, defined in such a way that it is intelligible to an inhabitant of a Topos, even if it is the relation we have just violently generated by a (set-theoretical) isomorphism between two inexistent sets… To reassure the purists, I would also point out that the relation $\leq$, which we have rather brusquely projected from 1 and $\subseteq$ toward the elements of C, can be reconstructed by purely categorial means. This relation will then be expressed as : '$f_1 \leq f_2$ if and only if there exists a unique arrow h such that the product-arrow $\langle f_1, f_2 \rangle$ is equal to the composition of h and the equalizer of the conjunctive (logical) arrow and the first projection of $C \times C$'. Doubtless the inhabitant of the Topos understands this more quickly than we do. And if we give him the diagram:

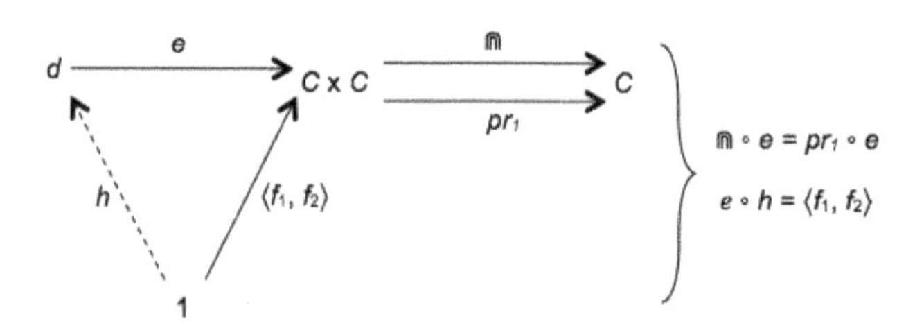

he will certainly be more capable of perceiving the consequences.
As for the function *J*, if in fact it is not an arrow of the Topos as such, it is nevertheless a *functor* (we shall see what this is presently) between the sub-category of sub-objects of 1 and the sub-category of the elements of *C*. We can assign to this a categorial meaning.

The conclusion to all this is simple: the study of an algebraic structure on the elements of *C* (and therefore of all that is required to make a Topos a logical place, a general 'evaluator' for the sum of statements) can also be presented as a study *of the sub-objects of 1*, and therefore of the possible operations on monomorphisms of the kind $a \rightarrowtail 1$, which also have the great advantage of being unique (once the source object is established), and therefore of being identifiable by an object (once *a* is given, the arrow $a \dashrightarrow 1$ is completely determined). So it is entirely within the categorial spirit to say something like 'the sub-object *a* (implied by 1)', without bothering to know if we're speaking of the arrow $a \dashrightarrow 1$ (which is more orthodox) or of the object *a* (which is a shortcut even the inhabitant of the Topos can allow).

21 INTERNAL ALGEBRA OF 1

The last chapter showed that the key to a Topos as a logical place resides in the (relational and operational) structure of sub-objects of 1, or of monomorphisms of the type $a \rightarrowtail 1$.

> *Note.* Here we observed how an arrow of the type $a \rightarrowtail 1$ is determined (as unique) as soon as a is fixed.
>
> We will therefore not hesitate to speak of the *sub-object a of 1*. This always refers to a monomorphism of source a and target 1, taken as representative of an equivalence class.
>
> We shall return to the convention of writing $1(a)$ whenever it is necessary to designate an arrow (especially when we mark the commutation of a diagram).

We will now define on these monomorphisms one relation (of order) and three operations.

a The (partial) order relation will be the one that we defined in the preceding chapter: $a \subseteq b$ if an only if there exists an arrow h (from the object a to the object b) such that $1(b) \circ h = \mathbf{1}(a)$.

b There will be a 'unary' operation, which makes correspond every sub-object a to a sub-object noted $-a$, which we name the complement of a. This complement will be associated to the arrow of negation, though the function J that projects the structure of 1 onto the elements of C (thus onto the logical arrows).

Note. The complement $-a$ is itself also an arrow going to 1. It is the name of an arrow which we write $-a \dashrightarrow 1$.

c There will be an initial binary operation which associates to two sub-objects a and b a third sub-object, c, an operation which we will call the *intersection* of a and b. We write this: $a \cap b = c$. The intersection of a and b finds itself associated through J to the conjunctive arrow $\Cap$ (which we have defined for every Topos), when it is applied to the product arrow $c(1(a)), c(1(b))$. In other words:

$$a \cap b = c \leftrightarrow \Cap \circ \langle J(a), J(b) \rangle = J(c)$$

remembering that (for example) $J(a) = c(1(a))$, and that the centration of an arrow of target 1, being an arrow of source 1 and target C, is really an element of C, thus a logical arrow (or an evaluating action).
The conjunctive arrow is in turn used to interpret, in any Topos, the conjunctive logical operator 'and', as in 'p and q'.

d There will be a second binary operation which we will call union of a and b. We write this $a \cup b = c$. Union, projected by J onto the elements of C, will here *define* the conjunctive arrow (which we haven't directly defined, but which is possible or even indispensible if we address ourselves to an inhabitant of a Topos!), noted $\Cup$, by posing that:

$$a \cup b = c \leftrightarrow \Cup \circ \langle J(a), J(b) \rangle = J(c).$$

This conjunctive arrow interprets, in any Topos, the disjunctive logical operator 'or' of statements, as in 'p or q'.

Note. We are not going to talk about what leads, through the algebra of 1 then its projection by J onto the elements of the Central Object, to the interpretation of the logical connective 'implication'. This is a very important but technical and rather delicate question that we will return to later. It suffices for now to know that we *can* define on the sub-objects of 1 an operation that, via J, interprets in any Topos the logical connective $\rightarrow$. And also (for an inhabitant of the Topos…) that we can directly define the arrow in question as a categorial combinatory operation (i.e. an operation between arrows) upon the elements of the Central Object.

a) Partial order on 1

Let's first note a basic feature of this order: two sub-objects of 1, say $1(a)$ and $1(b)$, are *always* comparable if an arrow h exists between a and b, or between b and a. In other words, any arrow of this kind commutes the diagram: we necessarily have, by the mere existence of such an arrow, $1(a) \circ h = 1(b)$ (which gives $1(b) \subseteq 1(a)$), or $1(b) \circ h = 1(a)$ (which gives $1(a) \subseteq 1(b)$).

Consider in fact the following diagram:

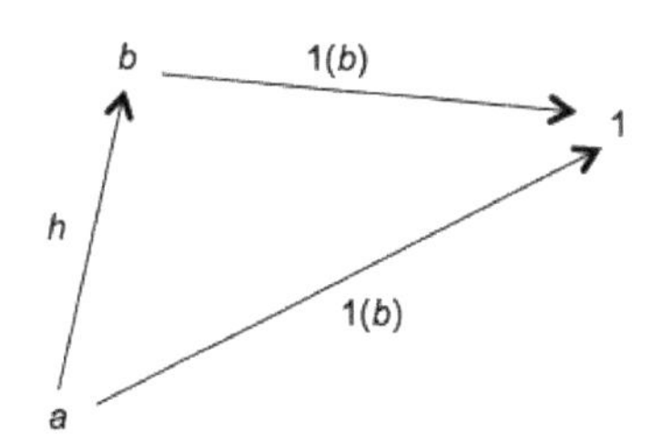

The arrow $1(b) \circ h$ is an arrow from a to 1. Yet there is only one arrow from a to 1 (since 1 is a terminal object), which is precisely $1(a)$. Thus the diagram commutes, and we have $1(a) \subseteq 1(b)$. The same reasoning shows that if there is an arrow h from b to a, we will have $1(b) \subseteq 1(a)$.

Amongst the sub-objects of 1, *the existence of an order-relation between 1(a) and 1(b) is strictly identical to the existence of an arrow between a and b, or b and a.* The sub-objects a and b are only incomparable with respect to the order when no arrow exists between a and b (thus categorially disjointed objects, or unrelated objects).

Philosophically, this means that two sub-objects of 1 are ordered whenever they are connected by an action. In other words, every action orders. This is the 'highly structured' side of 1, where every action operating from one of its sub-objects to another is immediately folded into the (partial) order established therein.

Let us add that if there is an arrow h from a to b and an arrow k from b to a, k is necessarily the inverse of h, h is an isomorphism, and a and b are similar monomorphisms, which therefore name the same sub-object.

Let's now examine a very special sub-object of 1, the arrow $0 \rightarrowtail 1$. The following simple diagram

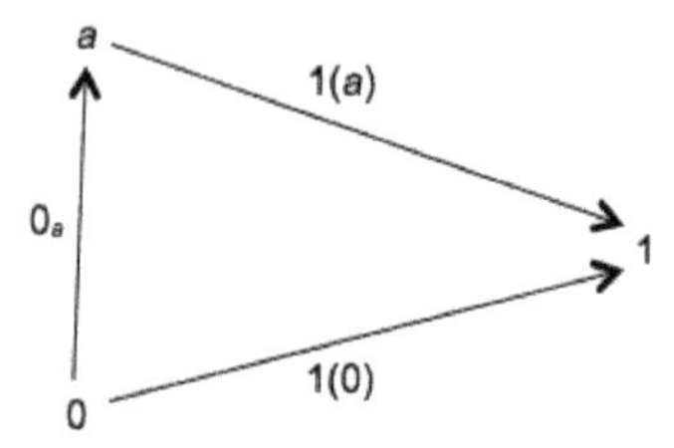

shows that we always have $1(0) \subseteq 1(a)$. Because the arrow $1(a) \circ 0_{a,}$ is an arrow from 0 to 1, and as such must be the arrow $1(0)$. Thus the diagram commutes, and we have $1(0) \subseteq 1(a)$.

Thus, the arrow $1(0)$ is itself, amongst the sub-objects of 1, and by the order-relation we are examining, 'inferior' to every other sub-object a.

We shall say that the sub-object of 1, which is 0, is a *minimum* for the order-relation.

Now let's examine the equally special sub-object of 1, which is the arrow $1 \rightarrowtail 1$, or $\mathbf{Id}(1)$. The diagram below

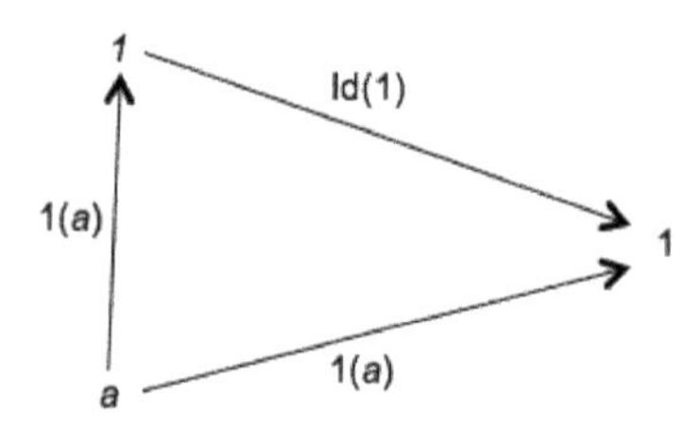

shows that we always have $1(a) \subseteq \mathbf{Id}(1)$. It is enough to note that every arrow in this diagram necessarily exists, and that $\mathbf{Id}(1) \circ 1(a) = 1(a)$ to see that the conclusion is inescapable. We will say the sub-object 1 of 1 is a *maximum* for the order-relation.

In sum: a partial order-relation exists on the sub-objects of 1, which comes into effect between a and b as soon as an arrow exists between a and b, and which has a minimum 0 and a maximum 1.

b) The intersection of two sub-objects

Take two sub-objects of 1, say a and b. Consider the pullback of a and of b (this always exists since the arrows $1(a)$ and $1(b)$ have the same target 1). Let's make $a \cap b$ the pullback-point, and $1(a) \cap 1(b)$ the *diagonal going from the pullback-point to 1*. We then have the diagram:

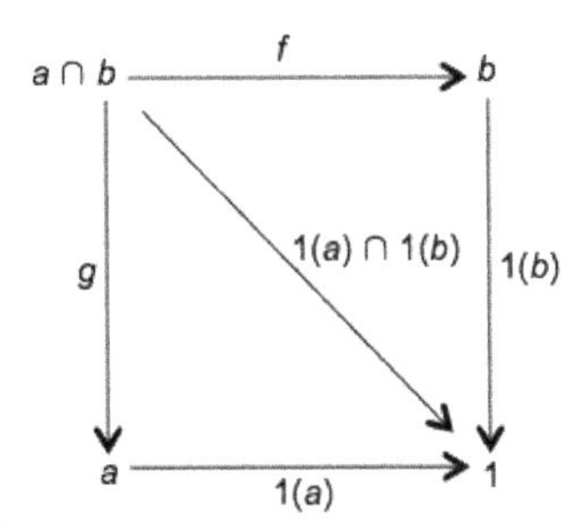

The arrow $1(a) \cap 1(b)$ will be named the *intersection* of $1(a)$ and $1(b)$. We will establish two essential propositions about the intersection.

Proposition 1: $1(a) \cap 1(b)$ is inferior (for the relation $\subseteq$) to $1(a)$ and to $1(b)$.

In the above diagram which defines intersection, the arrow $1(b) \circ f$ in an arrow from $a \cap b$ to 1. There is only one such arrow, which is simply the arrow $1(a) \cap 1(b)$. We therefore necessarily have $1(b) \circ f = 1(a) \cap 1(b)$ and consequently $(1(a) \cap 1(b)) \subseteq 1(b)$.

The same reasoning applies to the left triangle and to the composition $1(a) \circ g$, showing that $1(a) \cap 1(b)$ is inferior to $1(a)$.

Proposition 2: $1(a) \cap 1(b)$ is the largest (for the relation $\subseteq$) of the sub-objects that are smaller than $1(a)$ *and* $1(b)$.

In other words: if a sub-object c of 1 is such that $c \subseteq a$ and $c \subseteq b$, then $c \subseteq (1(a) \cap 1(b))$.

If a monomorphism $1(c)$ is inferior to $1(a)$, there exists an arrow h from c to a such that $1(a) \circ h = 1(c)$ (definition of order). And if the same $1(c)$ is also inferior to $1(b)$, an arrow k from c to b exists such that $1(b \circ k = 1(c)$. But therefore, we have $1(a) \circ h = 1(b) \circ k$.

Let us then extend the arrows h (from c to a) and k (from c to b) along the pullback which defines intersection.

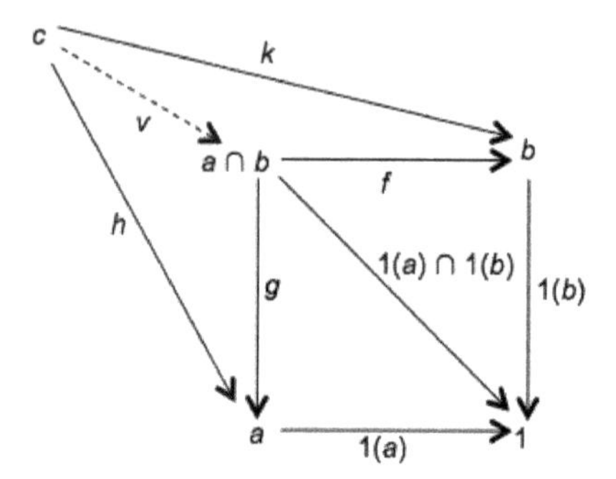

The exterior 'square' commutes, as a result of the equation above ($1(a) \circ h = 1(b) \circ k$). And consequently, there exists an arrow v from c to $a \cap b$ (the unique arrow which commutes the whole diagram by virtue of the pullback's limit function).

But the composition $(1(a) \cap 1(b) \circ v$ is an arrow from c to 1, which must therefore be the arrow $1(c)$. Thus we have $1(a) \cap 1(b) \circ v = 1(c)$, which, by the definition of order, means that $1(c) \subseteq 1(a) \cap 1(b)$, as we sought to demonstrate.

This proves, in the final analysis, that *the intersection of two sub-objects of 1 is the largest of the sub-objects of 1 simultaneously inferior to a and to b*. Or, if you prefer, 'the largest of the smallest (with regard to a and b)'. We will call $a \cap b$ the GI(a, b) (for 'greatest inferior of a and b') of the sub-objects a and b.

So the operation GI is defined upon the sub-objects of 1, which is in fact a feature of the order-structure: it assigns to any two sub-objects their intersection as GI(a, b). We can also say that the order-structure on the sub-objects of 1, which we have already seen admits the minimum 0 and the maximum 1, is also an order 'with GI'.

c) The complement of a sub-object of 1

The idea of the complement of a sub-object of 1 refers directly to centration, and in a way that anticipates the function of negation: the complement $-a$ of a sub-object a will in effect be *the sub-object of 1 whose centration is the negation of the centration of a*. That is to say:

$$c(-a) = \backsim \circ c(a)$$

Furthermore, it is impossible to speak of 'the negation of a', since negation is an arrow going from C to C, and which therefore cannot be composed with an arrow of the kind a···►1.

Negation is conceived as a logical arrow, thus as an arrow essentially applicable to evaluative actions: the elements of C.

But as every centration of a sub-object of 1 is precisely a logical arrow (from 1 to C), negation applies here, and will give the centration of the complement of a.

The following is a complete diagram of the definition of $-a$ (in two stages):

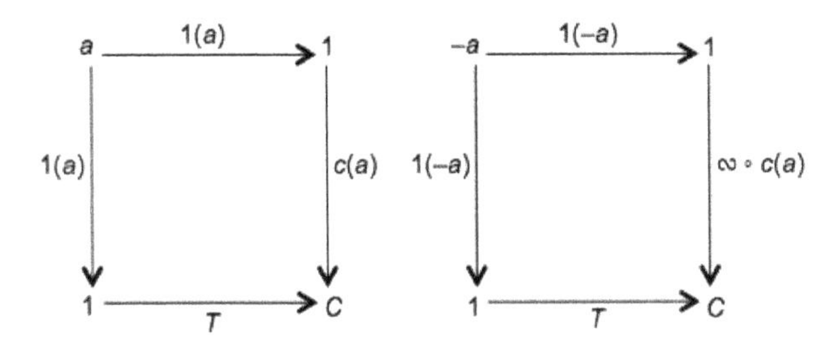

It could be argued that the arrow $-a$ is the 'indirect' negation of the arrow a: negation mediated by centration, which we know is like the logical analysis of monomorphisms, an analysis which, in the case where there is a sub-object of 1, is necessarily an element of the Central Object.

The definition of $-a$ enables us to know how the function J (between sub-objects of 1 and elements of C) works with regard to negation, insofar as we have:

1	$J(a) = c(a)$	(definition of J)
2	$c(-a) = \backsim \circ c(a)$	(definition of $-a$)
3	$J(-a) = \backsim \circ c(a)$	(by 1. and 2.)
4	$J(-a) = \backsim \circ J(a)$	(by 1. and 3.)

We can therefore say that J establishes an exact correspondence between the unary operation by which we pass from a to $-a$, or the complementary operation, and the operation of negation applied to elements of C: if you wish to know the value of J corresponding to the sub-object of 1 called $-a$, all you have to do is compose negation with the value of J corresponding to the sub-object a.

In brief: the unary operation of the complement *is* (according to the 'isomorphism' which J – set-theoretically – is), in the set of sub-objects of 1, the representative of negation.

The fundamental proposition of the complement is the following:

Proposition: The intersection of a and its complement $-a$ is similar to 0 (the arrow 0···►1, which is, I repeat, the *minimum* within the partial order of sub-objects of 1). In other words:

$$1(a) \cap 1(-a) = 1(0)$$

The elegant demonstration of this point is typical of categorial 'diagrammatic vision', and I urge you to follow of the details.

Consider the following diagram:

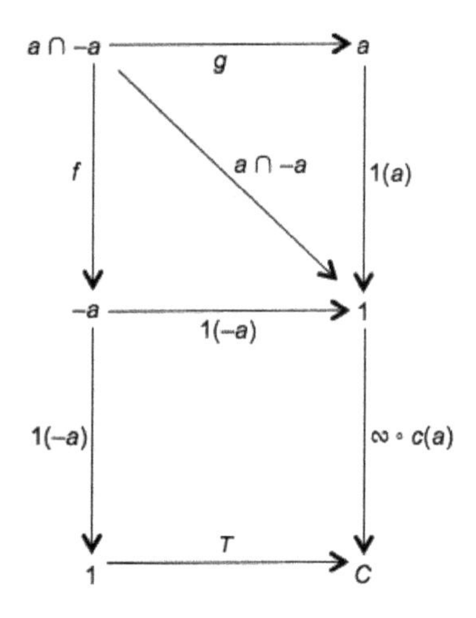

The top square is the pullback that defines the intersection $a \cap -a$. The bottom square is the pullback which defines the centration of $-a$ as being the negation of the centration of a. Therefore, by virtue of the famous 'pullback lemma', the complete rectangle is a pullback. So let's 'square' it:

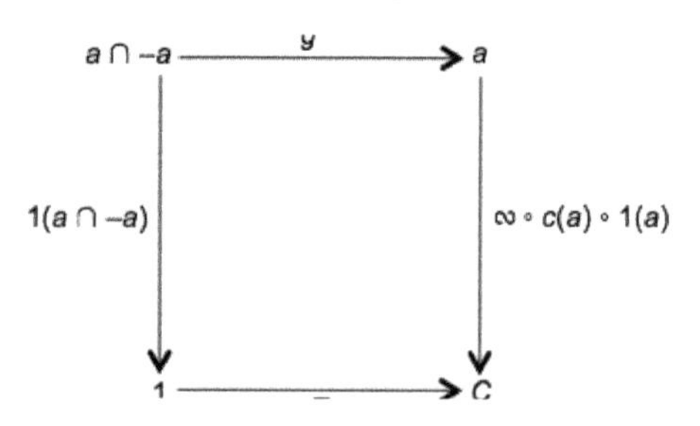

The right side is equal to $\infty \circ c(1(a)) \circ 1(a)$. But the (commutative) pullback which defines $c(1(a))$ presents itself thusly:

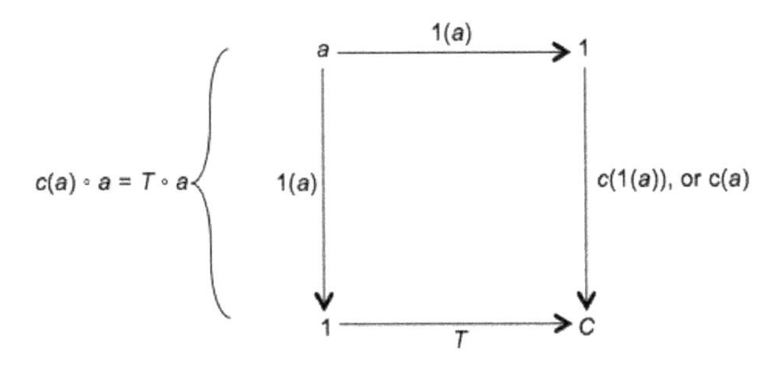

It then follows:

1 $\infty \circ c(1(a)) \circ 1(a) = \infty \circ T \circ 1(a)$ (the above pullback)

2 $\infty \circ T = F$ (property of negation)

3 $\infty \circ c(1(a)) \circ 1(a) = F \circ 1(a)$ (by 1. and 2.)

Finally, our 'squared' rectangle presents itself as follows:

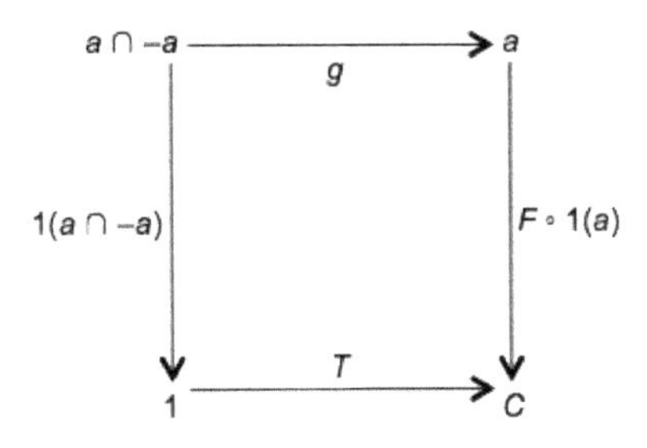

Its commutation requires that:

4 $F \circ 1(a) \circ g = T \circ 1(a \cap -a)$.

We will note that $1(a) \circ g$ and $1(a \cap -a)$ are two arrows from $a \cap -a$ to 1.

Let's extend these two arrows along the pullback which defines the false (recall that the false is the centration of the arrow $1(0)$):

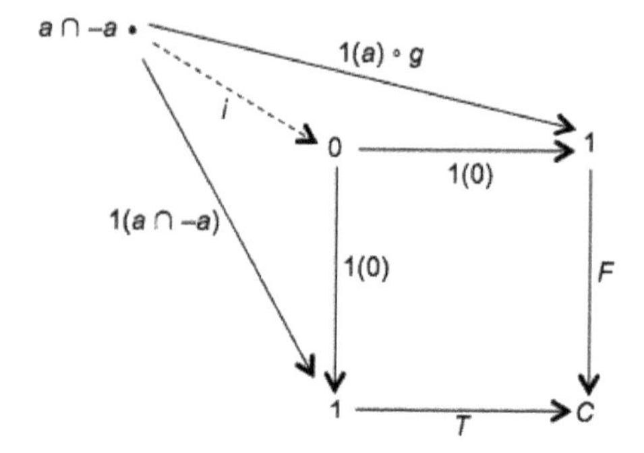

Equation 4. above expresses the commutation of the exterior 'square'. Consequently, there must be a (unique) arrow i from $a \cap -a$ to 0 which commutes the entire diagram (pullback as limit).

In particular, we have $1(0) \circ i = 1(a \cap -a)$, which, after the definition of the order-relation, signifies that

$$1(a \cap -a) \subseteq 1(0)$$

But we have also established that $1(0)$ was the minimum sub-object for the relation $\subseteq$. Thus the sub-object $1(a \cap -a)$ cannot strictly be smaller than $1(0)$, and the relation $1(a \cap -a) \subseteq 1(0)$ necessarily implies that $1(a \cap -a)$ is in fact the same sub-object as $1(0)$, and therefore that $1(a \cap -a)$ is similar to $1(0)$.

In short, we now have:

$$a \cap -a \simeq 0.$$

We showed earlier that $a \cap b$ was (for the partial order $\subseteq$) the largest of all the sub-objects inferior to a and b. We called this position the GI (greatest inferior) of a and of b.

Note that this proposition can therefore also be written algebraically, relative to the operation GI (if we return to monomorphisms):

$$\text{GI}(a, -a) \simeq 0$$

This evidently means that the *only* sub-object of 1 which is simultaneously inferior to a and to $-a$ is 0, which, being the minimum for the order-relation, is in any case inferior to everything.

This fundamental connection between the intersection and the complement is projected 'naturally' by J onto logical arrows.

1	$J(a \cap -a) = c(a \cap -a)$	(definition of J)
2	$c(a \cap -a) = c(1(0))$	(result of the above)
3	$c(1(0)) =$	(definition of the false)
4	$J(a \cap -a) = F$	(by 1., 2., 3.)

Note. These equations pass from *similarity*, when we operate on the monomorphisms of target 1, to *equality*, when we operate on the correlates of these monic arrows established through J (thus by centration). This is the whole point of the fundamental theorem of the Centre: to two similar sub-objects there corresponds (through centration) a unique element of C.

Furthermore, we know that the projection by J onto the evaluative actions (or logical arrows) associates an 'isomorphic' operation on the elements of C to every operation on the sub-objects of 1, and in particular that the conjunctive arrow $\Cap$ is associated through J to the intersection (which hasn't been demonstrated here, but is no less true). So we have:

5 $J(a \cap -a) = \Cap \circ \langle J(a), J(-a) \rangle$

By 5. and 4., the definition of J and the definition of $-a$ ultimately follows the fundamental relation on the elements of C:

$$\Cap \circ \langle c(a), ∾ \circ c(a) \rangle = F.$$

This essentially means that, irrespective of the evaluation in a Topos (on the basis of an elementary arrow of the Central Object) of the statement p, the evaluation of the compound statement 'p and not-p' will always be the false arrow.

Validation of the familiar law of the logic of statements.

From the fact that

$$\Cap \circ \langle c(a), ∾ \circ c(a) \rangle = F$$

it also follows that:

$$∾ \circ \Cap \circ \langle c(a), ∾ \circ c(a) \rangle = T$$

What this means this time is that irrespective of the evaluation of the statement p in a Topos, the evaluation of the compound statement 'not-(p and not-p)' always has the value T.

We will thus recognize that *every Topos evaluates the principle of non-contradiction (the impossibility of simultaneously affirming p and not-p) through the truth arrow.*

A Topos, even if it is absolutely non-classical (for example if it contains an infinity of distinct evaluative nuances and differences between the true and the false), determines the principal of non-contradiction as true.

Whence it turns out that the being of a Topos is established as consistent by the very thing that, for Aristotle, serves as the starting point of the science of being *qua* being.

d) The union of two sub-objects of 1

Let's launch straight into the definition: *The union of two sub-objects a and b of 1 is the image of the co-product arrow of the arrows a and b.* Namely:

$$a \cup b = \text{im}([a, b])$$

The co-product is the dual of the product (cf. §5).

We defined the image of an arrow in §19 (theorem 7): it is the monomorphism which is part of the canonical decomposition of an arrow (of a Topos) into an epimorphism followed by a monomorphism.

It follows that the two diagrams below demonstrate the function of the union $a \cup b$ (we note that this is indeed an arrow of target 1, thus a sub-object of 1).

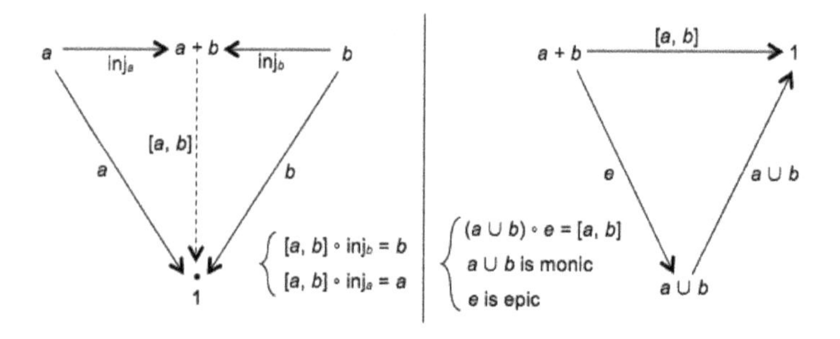

Note. The definition of union is obviously far less transparent than that of the intersection or the complement. In general, the categorial presentation of the arrows which relate (interpret) the logical operator of disjunction (namely, 'or' as in 'p or q') is very intricate and hardly 'speaks for itself'.

But what this complication reveals *is the stark contrast (for thought as well) between the conjunction 'and' and the disjunction 'or'.*

Conjunction is basically a symmetrical connective: 'p and q' is true if p and q are both true. Every dissymmetrical case for the conjunction of p and q (false p and true q, false q and true p) gives the false.

It is not at all the same for disjunction. In the classical interpretation, disjunction is true in the symmetrical case (true p and true q), but also in every dissymmetrical case. It is only false if p and q are both false.

This 'unstable' property produces all manner of chicanery in thought, which Lacan takes full advantage of in his celebrated interpretations of the saying 'your money or your life!'[39]

Basically, the 'open' logical spaces of Topoi once again expose the root of this disjunctive chicanery, relative to which conjunction functions with a uniform simplicity.

This root is that *the algebra of the sub-objects of 1 is largely the same for all Topoi when it comes to the operation GI (greatest inferior), which is founded on the intersection, but when it comes to the operation SS (smallest superior), which we will define from the union, the situation is different*. In fact, the characteristics of the operation SS, thus of union, defines two large types of Topos corresponding to universes whose immanent logic is not the same: classical on the one hand, intuitionistic or sub-intuitionistic on the other.

At first, however, the investigation of the union's properties is absolutely symmetrical to that of the intersection. Just as we showed that the intersection of two sub-objects of 1, say a and b, is the largest of the sub-objects which are inferior to a and to b (operation GI), we will equally show that the union of these two sub-objects is the smallest of the sub-objects which are superior to a and b (operation SS for 'smallest superior'.)

Note. The use made here of terms 'superior', 'inferior', 'smallest' or 'largest' refers to the order structure defined on the sub-objects of 1, and noted $\subseteq$.

The metaphorical expression 'a is inferior to b' or 'a is smaller than b' must be subtracted from any thought of 'size'. After all, a and b are only arrows of a Topos, and we are far from having defined any measure on these arrows.

In truth, 'a is inferior to b' strictly means that a stands to the left in formula $a \subseteq b$, just as 'b is larger than a' means that b stands to the right. The metaphor is based on the fact that $\subseteq$ is a reflexive and transitive order-relation, and our immediate linguistic intuition of such a relation is effectively associated with 'larger' and 'smaller'. But in the case of the relation on the sub-objects of 1, it is necessary to see that these metaphors must be referred back to pure algebra.

Moreover, as the relation $\subseteq$ is reflexive, we find that we have things like $b \subseteq b$. We must therefore say that b is at the same time larger and smaller than itself, superior and inferior to itself, at which point the limits of the metaphor become apparent. The most accurate reading of the relation $a \subseteq b$ is: 'b is superior *or equal* to a'.

Care should be taken when handling conveniences like 'superior' or 'lesser than' that these assertions of relative position always incorporate equality as a possible case (where 'equal' means: simultaneously inferior and superior).

We have already encountered the intra-categorial expression of this point: if we have at the same time $a \subseteq b$ and $b \subseteq a$, it is because – by the definition of order – there exists an arrow h from a to b which commutes the relevant diagram, and an arrow k from b to a which does the same. It is then obvious that k is the inverse of h, that h is an isomorphism which commutes the diagram, and therefore that the monomorphisms $1(a)$ and $1(b)$ are similar, meaning they designate *the same sub-object*.

As with the intersection $\cap$, the investigation of the union $\cup$ takes the form of two propositions.

Proposition 1: Two sub-objects a and b are both inferior to their union. So we have $a \subseteq a \cup b$, and $b \subseteq a \cup b$.

Let's apply to the diagram defining the co-product of $1(a)$ and $1(b)$ (see below) the epic-monic decomposition of the co-product arrow $[a, b]$, where $a \cup b$ is the monic part, or the image of $[a, b]$ (idem). We obtain the consolidated diagram below.

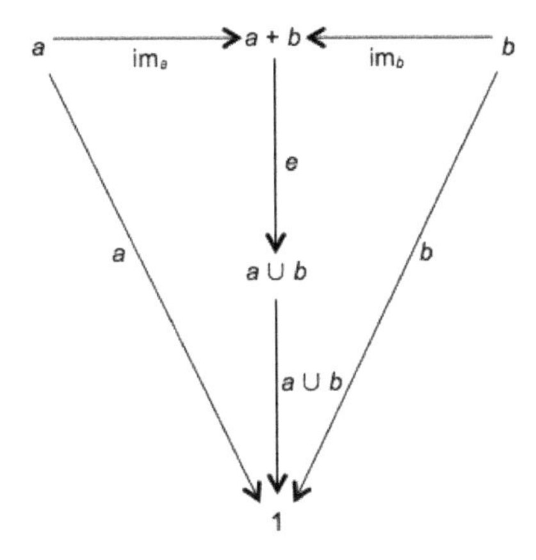

As the diagram commutes, it follows that:

1	$a \cup b \circ e \circ \text{im}_b = 1(b)$	(commutation)
2	$a \cup b \circ (e \circ \text{im}_b) = 1(b)$	(associativity)
3	$1(b) \subseteq a \cup b$	(definition of order)

Similarly, by 'treating' the commutativity of the left triangle, we demonstrate that $1(a) \subseteq a \cup b$.

Proposition 2: The union $a \cup b$ is the smallest of the sub-objects of 1 that are simultaneously superior to both a and b.

We will say that $a \cup b$ is the smallest of the superiors, noted SS, and we will write:

$$a \cup b = \text{SS}(a, b)$$

While the demonstration of this point is markedly intricate, so we will not present it here. We are in any case in possession of an operation SS(a, b) on the sub-objects of 1, and everything seems to indicate that it symmetrizes the operation LI(a, b). We say that any two ordinary sub-objects of 1 are 'framed' by their GI ('above') and SS ('below').

To these operations on the sub-objects of 1 correspond, by the (set-theoretical!) isomorphism J, operations on the evaluative actions or elements of the Central Object. For the GI, there corresponds the arrow ⋒, which we have directly defined as centration of the product-arrow $\langle T, T \rangle$, and for the SS, it is the arrow $\Cup$, which we have not directly defined (to the chagrin of the inhabitant of the Topos, who is baffled as to why this is the case, given that is that it is *possible* to define it). This definition would be: 'centration of the image of the co-product arrow $C + C$ toward the product $C \times C$'; the said arrow being itself the co-product of two arrows, which in turn are the product $\langle T, \mathbf{Id}(C)\rangle$, and the product $\langle \mathbf{Id}(C), T\rangle$!

These operations on the elements of C interpret in turn, in any Topos, the two great logical connectives of conjunction (p and q) and disjunction (p or q), by taking as possible values of statements p and q elements of C.

We have seen in particular that the crucial theorem of the intersection of two sub-objects, namely $1(a) \cap 1(-a) = 1(0)$, was the interpretation in a Topos, via the 'isomorphic' operation on the elements of C, of the logical law: 'p and not-p is always false'.

One would expect, given the symmetry hitherto maintained, to have a dual theorem which affirms that, in every Topos, $1(a) \cup 1(-a) = 1(1)$. This would mean that, just as the intersection of a sub-object and its complement is the minimum-arrow 0, so too the union of a sub-object and its complement is the maximum-arrow 1. In other words, that vis-à-vis:

$$\mathrm{GI}(a, -a) = 0$$

we have:

$$\mathrm{SS}(a, -a) = 1$$

However, and this fact is absolutely crucial, *this is not the case.*

It is not true *in every Topos* that the union of the sub-object of 1 and its complement is equal to 1. There are Topoi where this is true and Topoi where this is not true.

Let's say that the statement 'the union of a sub-object and its complement is equal to the maximum sub-object $\mathbf{Id}(1)$' is, strictly from the viewpoint of the general definition of a Topos, *undecidable.*

But what does this statement mean logically?

Through the 'isomorphism' J, this statement is transferred to the elements of the Central Object in the following way:

- The arrow $\Cup \circ \langle c(a), c(b)\rangle$ corresponds to the union of a and b
- But if $b = -a$ then we have, by the definition of the complement of an object, $c(b) = c(-a) = \infty \circ c(a)$.

- And finally, the centration of the maximum sub-object **Id**(1) corresponds to the centration of that object. Now we know that the centration of **Id**(1) is T, the true.

We have therefore, for any Topos where it would be true that $a \cup b = 1$, the following equally true statement:

$$\Cup \circ \langle c(a), \backsim \circ c(a) \rangle = T$$

Now we know that the arrow $\Cup$ interprets in a Topos (through the elements of C) the logical connective of disjunction, namely, 'or', and that the arrow $\backsim$ interprets negation. We know also that the arrow T interprets the value of truth, whose name it bears.

The above statement thus interprets: 'p or not-p is true'.

Yet 'p or not-p' is nothing other than the *famous principle of excluded middle*, which, among other things, is the basis for reasoning by the absurd.

Just as $a \cap -a = 0$ interprets the principle of non-contradiction in a Topos, so too the statement $a \cup -a = 1$ interprets the excluded middle.

That one is valid in any Topos and the other undecidable is a very striking dissymmetry.

The general notion of universe, crystallized by the concept of Topos, universally validates the principle of non-contradiction (on the basis of the theorem GI(a, –a) = 0), but leaves the principle of excluded middle hanging (no theorem SS(a, –a) = 1).

Everything happens therefore as if this concept of universe, conforming to the exploration of possible logics that it proposes, simultaneously renders possible the admission of the 'classical' principle of excluded middle, and its 'intuitionistic' rejection.

And it is in fact just as compatible with the general concept of Topos to admit this principle as it is to reject it.

So here we have reached a point of bifurcation: there are two fundamental 'kinds' of universe (or Topoi):

- Those which validate the principle of excluded middle on the basis of the legitimacy, in the sub-objects of 1, of the equation: $a \cup -a = 1$. This Topos will be called *classical.*
- Those which do not validate this principle, since the equation $a \cup -a = 1$ is not applicable here to every sub-object of 1. This Topos will be called *non-classical.*

The categorial concept of universe lies *beyond* the opposition between classical and non-classical logic. It exhibits a consistency that leaves this opposition undecidable. The decision on this point takes us into the plurality of universes, into the kinds of logic that its generality subsumes.

22 ONTOLOGY OF THE VOID AND EXCLUDED MIDDLE

Here we will establish one of those onto-logical correlations that give Category Theory its charm. It knots an ontological characterization of universes (the uniqueness of the void) to a logical prescription: the validity of excluded middle, of the statement 'p or not-p'.

Clearly, set theory ontology is founded on the uniqueness of the referent of the name of the void, *and* validates the excluded middle: it makes dramatic use of reasoning by the absurd, which is sustained only by the excluded middle (one uses the impossibility of not-p to *impose*, without directly demonstrating, the truth of p). But set theory fails to think the correlation between the two, or at any rate it does not think it as singularity, or as a possible law of thought. The force of the categorial investigation is in establishing that between the ontology of the void and the validation (or non-validation) of excluded middle there is a relation that is *a law of (possible) universes.*

We see once again that the categorial way of thinking, arranging the possible universes through the movement that goes from logical plurality to ontological singularity, *clarifies* the vital points of the ontological decision itself.

The onto-logical link here takes the form of the following theorem, whose philosophical importance is evident:

Theorem: In any Topos where the 'void' and 'zero' coincide (where every void or element-less object is isomorphic to zero, or initial), the equation $a \cup -a = 1$ is demonstrable. Therefore this Topos validates the excluded middle.

We will also say: in any universe where the void is unique, the logic is classical.

By contrast, a non-classical Topos necessarily admits 'multiple' voids, which is to say it presents element-less objects which are not isomorphic to 0 (which are not initial).

Take a Topos where every void object is isomorphic to the initial object 0. Now take a sub-object a of the terminal object 1. Consider the union of a and its complement, namely, $a \cup -a$.

Two cases then present themselves:

a) $a \cup -a$ is isomorphic to 0.

Note. Confusion is again mounting! In virtually the same phrase we use 0 to designate:

- (First use): the initial object of the Topos.
- (Second use): the minimal sub-object of 1, which is the arrow 1(0), or $0 \dashrightarrow 1$, the arrow that has the first 0 as its source!

But this is only one example of the abuse of language that we observed above. We pass from the arrow 1(0), which is entirely determined by its source 0, to the marking of this arrow by 1, and finally to the designation of a sub-object of 1 (the collection of monomorphisms similar to the arrow 1(0)) by the same letter.

The categoricians are not so easily offended. One of their 'natural' values is that of determining the multiple meanings of a single sign depending on the shifting context. As in 'natural' language the same word can take on many meanings depending on the context of the phrase or the page.

So, placing our faith in the context, we will not worry that 0 as the initial object of the Topos becomes 0 as the minimum sub-object in the order defined on the sub-objects of 1, reminding ourselves from time to time that the first is an object and the second an arrow. Just as 1 will moreover come to equivocate between the terminal object of the Topos and the arrow 1(1), or **Id**(1), which is the maximum sub-object in the algebra of 1.

If $a \cup -a$ is isomorphic to 0, then the arrow $1(a \cup -a)$ is similar to the arrow 1(0) (we are taking some care with the notation here). Consequently, the sub-object $a \cup -a$ is the same as the sub-object 0 (we are abandoning our precaution).

But, by virtue of the properties of union like SS, a and $-a$ are inferior to their union $a \cup -a$, and therefore to 0 as well. But this is only possible, since 0 is minimal, if they are both *similar* to 1(0) (if you consider them as monomorphisms), or equal to 1(0) (otherwise known as 0), if you consider them as sub-objects (cf. above: 'inferior' meaning 'inferior or equal').

In any case, a and $-a$ *have the same centration as 1(0)*, since the centrations of two similar arrows are equal.

Note: If we identify ourselves with the inhabitant of a Topos, we can see the equivocation of nominations more clearly. For the inhabitant, *there are only arrows and objects*, and while two isomorphic objects are literally distinct (as pure presentations), they are 'actively' (thus actually) indiscernible. 0 designates a terminal object, and thus if we *also* use 0 to designate the arrow $0 \rightarrow 1$ (where 1 is a terminal object), this is clearly a homonym bearing on two different kinds of reality.

What the inhabitant of a Topos sees is that 'all' the similar monorphisms of target 1 (which for him are different arrows) have the same centration. What he can't see is this 'all', he doesn't see it as One (it is neither an arrow nor an object). By contrast he can perfectly see that the centration of several (similar) arrows is *one* arrow from 1 to C.

We, hopelessly exterior set-theoretical ontologists, can conclude that what we call 'sub-object', and which is the One of a collection, is given as One in the interior of a Topos (for the inhabitant) only *in the form of the unique centration of distinct arrows*. From an immanent point of view, a 'sub-object' of 1 is a centration, thus an arrow from 1 to C.

So after all there will be – regarding names – a possible if regrettable homonymy between the name of the initial object 0 and that of the unique arrow $0 \dashrightarrow 1$. The same goes for something else as well, namely, the centration of the arrow $0 \dashrightarrow 1$, noted $c(1(0))$, that, if you like, One-ifies the arrows similar to 1(0), which could be many, since the Topos may be swarming with as many objects isomorphic to 0 as you want. We can then call this centration the 'sub-object of 1'. It would however be regrettable to keep naming this 'sub-object' 0, since it is an element of the Central Object, and thus this time an arrow whose *source* is 1. In fact, it is called F (the false).

We will add one last subtlety: what is it that, in a Topos, somehow 'represents' every arrow from 1 to C? It is, we know, the exponential object C^1. If a sub-object of 1, for the inhabitant of the Topos, is at best the centration of a monomorphism of target 1, the 'collection' of sub-objects is more or less indicated by C^1.

But we saw in §17, that C^1 is isomorphic to C, and therefore is, for the inhabitant of the Topos, an object which – from the perspective of the purely relational and operational values of the universe – is similar in every respect to C itself.

So that finally, for an inhabitant of the Topos, what we call the collection of sub-objects of 1 is in fact quite well represented… by the Central Object!

Which is an immanent version of our 'exterior' correlation J between the sub-objects of 1 and the elements of C.

Returning to our thread: in the case where $a \cup -a = 0$, a and $-a$ have the same centration as the arrow 1(0).

But the centration of the arrow 1(0) is our old acquaintance: the false arrow. It is even its definition.

Thus $c(a) = c(-a) = F$

Only, by the definition of the complement $-a$, we know that $c(-a) = \infty \circ c(a)$.

Replacing $c(a)$ and $c(-a)$ with their common identity, which is F, we then have:

$F = \infty \circ F = T$ (Cf. §15)

Yes, but if the false arrow is the same as the true, the Topos is degenerate (§15). *By contrast*, if the Topos is not degenerate (and this is, after all, the only case that matters to us), the union of a and of $-a$ cannot be isomorphic to 0 (or the initial object), since the supposition of this isomorphism leads us straight to $F = T$.

For a non-degenerate Topos, we have therefore necessarily the second case:

b) *a* ∪ *–a* is not isomorphic to the initial object 0.

It is here that we must remember our ontological hypothesis: the Topos is supposed to admit only a single void (up to isomorphism): Zero. So if $a \cup -a$ is not isomorphic to 0, then it is not void. Thus, it possesses an element x, which is a elementary arrow $1 \overset{x}{\dashrightarrow} (a \cup -a)$.

We then have the following diagram (note once again that by $a \cup -a$ we understand both the source object of the union of a and $-a$ and the union arrow itself):

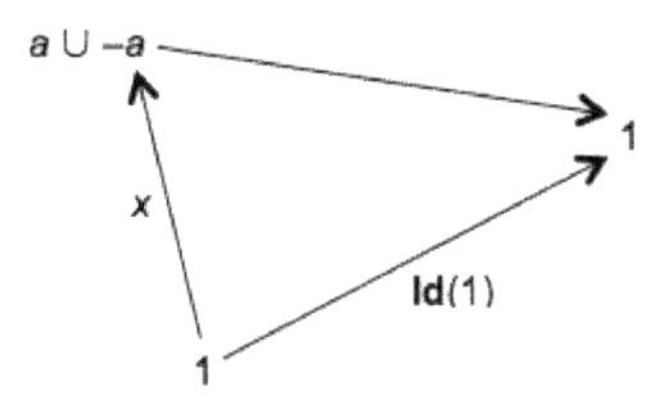

It is undoubtedly commutative, and by the definition of order on the sub-objects of 1, it follows that

$\mathbf{Id}(1) \subseteq 1(a \cup -a)$ (taking care with the notation)

or (taking less care now):

$1 \subseteq (a \cup -a)$

But we know that 1 is the maximum for order. So it also follows that:

$(a \cup -a) \subseteq 1$

These two equations are only simultaneously possible if $(a \cup -a) = 1$.

Consequently, if its not degenerate, a Topos which is ontologically marked by the uniqueness of the void (every void object is initial) positively decides the statement whereby the union of a sub-object of 1 and its complement is equal to the maximum 1 (or **Id**(1)).

This is to say that such a Topos validates the logical principle of excluded middle.

To go from the uniqueness of the void to the licit usage of reasoning by the absurd is the correct conclusion. Which proves that this reasoning presupposes an ontological decision, which is in short *a Parmenidean principle applied to the negation of Parmenides.*

The negation of Parmenides amounts to saying that the nothing of every presentation *is* (there is a void).

The Parmenidean principle amounts to saying that everything which is, in so far as it *is*, is One.

The combination of these two gives: insofar as it *is*, the void is One.

This immediately engages a universe which in fact validates reasoning by the absurd, since it choses, amongst possible universes (Topoi), one which validates the principle of excluded middle.

By way of conclusion let's note that we know that a well-pointed Topos (which admits a local marking of difference) tolerates only a single void (this is what is at stake in our first big ontological relation, cf. §18). Thus a well-pointed Topos necessarily validates the law of excluded middle.

Hence another general connection revealed by the extent of the categorial investigation: *if every difference allows itself be evaluated by a point, then excluded middle is a legitimate logical principle.*

We therefore see taking shape two *possible lines of thought*:

1 Unity (ontological, of the void), localization (of difference), classicism (of logic).

2 Plurality (of voids), globalization (of differences), intuitionism (of logic).

But then again, the general concept of possible universes (of Topos) subsumes these two lines, proposing a configuration where what separates them is left as an undecidable statement.

23 A MINIMAL CLASSICAL MODEL

We have established that a Topos's 'logical type' essentially depends on the algebraic structure of the sub-objects of 1, and in particular on the equation:

$$SS(a, -a) = 1$$

It is in fact this equation which controls the validity of the principle of excluded middle '*p* or not *p*' in the Topos.

It is therefore illuminating to isolate the algebraic structure on the sub-objects of 1, giving them (in a provisionally set-theoretical style) a formal allure, independent of any consideration of universe.

Let's summarize the pertinent traits (with regard to the pure determination of the immanent logic of Topos) of this algebra:

1. There is an order, which is generally a non-strict partial order (it admits $a \subseteq a$)
2. This order had a minimum 0 and a maximum 1.
3. To every element a (sub-object of 1) corresponds its complement, noted $-a$.
4. To every pair of elements b, c corresponds the LI(b, c), whose fundamental property is that it is the largest of the elements simultaneously inferior to b and c.
5. To every pair of elements b, c corresponds the SS(b, c), whose fundamental property is that it is the smallest of the elements simultaneously superior to b and c.

6. We have the equation: $\mathrm{LI}(b, -b) = 0$
   The equation $\mathrm{SS}(b, -b) = 1$ is not prescribed. It depends on the singularity of the structure (thus of the considerations of the universe).
   The minimal model [*miniature élémentaire*] corresponding to this description consists in retaining *only* the minimum 0 and the maximum 1 as elements (since they are obligatory), and defining on the set (0, 1) the relation $\leq$ which, in addition to its reflexive constraints ($0 \leq 0$ and $1 \leq 1$), requires that $0 \leq 1$. In short, the 'usual' order on the set (0, 1).
   We note that this order is total.
   It is obvious that 0 is minimum and 1 is maximum.

The operations have a quasi-obligatory definition, by virtue of two remarks:

1. In *every* structure of this kind, $\mathrm{LI}(a, a) = \mathrm{SS}(a, a) = a$

   Note. This is the price paid for the reflexivity of order, thus for the fact that we always have $a \subseteq a$. Cf. the above remarks on the fact that 'inferior' or 'superior' subsume equality. It follows that a is at the same time the largest element inferior to itself, and the smallest element superior to itself. Hence the above result. In algebra, we would say that the operations GI and SS are idempotent.

2. In a non-degenerate Topos, if $b = -b$, b can be equal to neither 0 nor 1.

   We used this point in our demonstration of the classical character of any Topos which upholds the uniqueness of the void. If $b = -b$, we also have the equality of centrations, namely, $c(b) = c(-b)$. But $c(-b) = \infty \circ c(b)$, by the definition of the complement. Furthermore, if b is 1 (the arrow $\mathbf{Id}(1)$), then $c(b)$ is the true arrow, and if b is 0 (the arrow 1(0)), then $c(b)$ is the false arrow. In any case, we end up with true = false, by the operation of the arrow of negation. In which case the Topos is degenerate.

   In order that our minimal model is adequate, it is therefore necessary that we have:

   $-\mathrm{LI}(1, 1) = \mathrm{SS}(1, 1) = 1$
   $-\mathrm{LI}(0, 0) = \mathrm{SS}(0, 0) = 0$
   $-0 = 1$
   $-1 = 0$

Moreover, as we must have LI(x, $-x$) for every value x (in this case, 0 and 1), it necessarily follows that $-$LI(1, 0) = LI(0, 1) = 0, which perfectly corresponds to the intuition of the 'usual' order upon (0, 1).

Trusting in this intuition (the SS must be superior to two elements of which it is the SS), we finally get:

$$-\text{SS}(1, 0) = \text{SS}(0, 1) = 1.$$

We can summarize all of this in three double-entry tables

complement		LI			SS		
x	−x		1	0		1	0
1	0	1	1	0	1	1	1
0	1	0	0	0	0	1	0

It is obvious that LI(x, $-x$) = 0 for every x (whether 0 or 1) (see for yourself).

An essential remark is that the equation interpreting the principle of excluded middle is demonstrable in this minimal model, since SS(x,$-x$), by the above remark which necessarily differentiates x from $-x$, is ultimately SS(1, 0) or SS(0, 1) and is therefore, according to the table, always equal to 1.

Our minimal model is therefore *classical.*

This should not surprise us, since our model is nothing other than the standard interpretation of the propositional calculus. It suffices to think of 1 as the true and 0 as the false: the above tables then becomes those 'truth tables' taught in school, where the complement corresponds to negation, the GI to conjunction (p and q), and the SS to disjunction (p or q).

To what, in a Topos, does this minimal model correspond? Can we 'erect' it in a universe?

The first remark to make is that if the algebra on the sub-objects of 1 is isomorphic to the minimal model (0, 1), it means *that there are only two sub-objects of 1*, corresponding to 0 and to 1. In other words, the only monomorphisms whose target is 1 (up to isomorphism) are necessarily the minimum (the arrow 1(0)) and the maximum (the arrow 1(1), or **Id**(1)).

But these sub-objects are in bi-univocal correspondence with the elements of the Central Object, via the function J, which is to say the centration. Therefore, there are only two elements of C also.

What are they? Obviously, the true arrow (centration of 1) and the false arrow (centration of 0).

Consequently, any Topos whatsoever whose algebra of sub-objects is (abstractly) our minimal model is a *bivalent Topos.*

The inverse is also correct. Take a bivalent Topos. If it is not degenerate, 1 is different to 0, and the true is different to the false. And we have seen above that $-0 = 1$, and $-1 = 0$ as a matter of course.

We also know that $\text{LI}(a, a) = \text{SS}(a, a) = a$

Which leaves the 'dissymmetrical' cases for the GI and the SS.

For the LI, the case is simple: as $-0 = 1$, and we necessarily have $\text{LI}(a, -a) = 0$, it follows that $\text{LI}(1, 0) = \text{LI}(0, 1) = 0$

For the SS, we draw from the demonstration of the theorem from §22, where we saw that if $a \cup -a = 0$, the Topos is degenerate. It is therefore clear that if it is not degenerate, and if it is (like ours) bivalent, then $a \cup -a = 1$ (there is no choice, if 0 is impossible). Which gives us $\text{SS}(1, 0) = \text{SS}(0, 1) = 1$

Finally, the table of operations ('complement', 'union' (or SS) and 'intersection' (or GI)) is exactly the same for the two sub-objects of 1 as it is in our minimal model. This model does in fact give us the formal schema of the algebra of sub-objects (for these three operations).

The obvious conclusion is that, for these three operations, *the logic of every bivalent Topos is classical.*

In particular, a bivalent Topos necessarily validates the principle of excluded middle.

This is another way of establishing the connection between the ontological property of difference and classicism of logic. Because we have already shown (§19, theorem 8) that every well-pointed Topos is bivalent, meaning any Topos which locally avers differences between arrows, validates excluded middle, and is classical with regard to the logical operators of negation, disjunction and conjunction.

24 A MINIMAL NON-CLASSICAL MODEL

Since we just established that every bivalent Topos validates the excluded middle, it is inconceivable to point to a non-classical model which only has two elements.

Hence the aphorism: The Two is classical.

Ever since Hegel – but doubtless more fundamentally since Christianity – we know that the overturning of classicism requires the Three.

So let's take three elements: 0 (minimum), 1 (maximum) and M (middle).

To consolidate these names, we will define the order-structure on the triplet (0, M, 1) in addition to the requisite reflexive relations of the kind M ≤ M etc. by the chain:

$$0 \leq M \leq 1$$

So the minimum is 0, the maximum is 1, and M is 'intermediate'.

What is our guide in the construction of the 'complementary' operations GI and SS? The will-to-not-be-classical. Suppose for at least one value of x (0, 1, or M), we have SS(x, –x) ≠ 1. This will mean that for the evaluation which confers on p the value x, the statement 'p or not p' is not valid, whence it follows that the excluded middle is not a universal logical law.

It is obviously the new element M that will hold us back.

In fact, we hardly have a choice (if our desire is non-classical).

We cannot assign to the complement –M of M the value M. Because, given the order displayed, LI(M, –M) = LI(M, M) = M (as noted in §23), which would contradict the fact that we have, in any Topos, LI(a, –a) = 0.

This doesn't mesh with our desire to assign to –M the maximum value 1. Because under these conditions, SS(M, –M) = SS(M, 1) = 1, since 1 is certainly the smallest superior to M and to 1. Although, as a result, the structure perfectly validates the excluded middle for the value M.

The only possibility left for us is:

$$-M = 0$$

And this, to tell the truth, is all we need to do. Because, relying on our intuition of order, we then see that SS(M, −M) = SS(M, 0) = M, since M is definitely the smallest of those that are bigger than – which is to say, superior or equal to – 0 and M.

This is where the principle of excluded middle falters.

We already know that the fate of 0 and of 1 is, by means of centration and the arrows 1(0) and 1(1), to refer to the true and false arrows, leaving very little room to move, since $T = \backsim \circ F$ and $F = \backsim \circ T$, and that negation $\backsim$ 'transfers' (through J via the evaluative arrows) the complementary operation onto the sub-objects of 1.

Without holding back, we will pose that −0 = 1 and −1 = 0, just like in the minimal classical model.

The remaining operations go without saying, and are summarized in the following table:

complement		LI				SS			
x	−x		0	M	1		0	M	1
0	1	0	0	0	0	0	0	M	1
M	0	M	0	M	M	M	M	M	1
1	0	1	0	M	1	1	1	1	1

That the principle LI(x, $-x$) = 0 is verified for the three values of x is trivial. In fact, the only novelty with regard to the minimal classical model concerns M, and we have LI(M, −M) = LI(M, 0) = 0 (0 is the only element inferior to 0).

Note. An important remark concerning *double negation.* Since the complement is tied to the negation arrow, we might expect that, if the logic is classical, the fact that complement is seized 'twice', as $-(-a)$, which 'equates' through centration to a double effect of negation (so $\backsim \circ \backsim \circ (c(a)))$, would bring us back to affirmation. It would be in the best classical taste to end up with $-(-a) = a$ for every sub-object of 1, and as a consequence $\backsim \circ \backsim \circ (c(a)) = c(a)$ for the elements of C. The last equation would be the evaluation in a Topos of the great law of classical logic 'not-not-p is equivalent to p'.

This is indeed what happens with our minimal classical model, and the corresponding bivalent Topoi. Since – for example – we have $-(-1) = -(0) = 1$.

We will say that this Topos validates the law of double negation.

But this is not what happens in our minimal non-classical model. Because in the case of M, we have $-(-M) = -(0) = 1$. And therefore, as the complement interprets negation, it must be admitted that the negation of the negation of M does not give back M (it gives 1).

This is another crucial definition of non-classical evaluations: they don't admit a universal validation of 'not-not-p equating to p'. We know, moreover, that the principle of double negation is itself also central to the contention between the classical and the intuitionistic orientations.

The meaning in a Topos of this bifurcation takes on more and more importance, above all when we see that in a non-classical Topos, double negation functions *as a topological operator*. The basic idea being that the negation of the negation of an arrow, while it is not identical to this arrow, is 'infinitesimally' close.

Our minimal classical model directly referred to the standard evaluation of the logic of statements. To what style of interpretation does our minimal non-classical model refer?

Of especial interest is what is called, after the name of its inventor, Kripkean semantics.

Entirely beholden to intuitionism, this kind of evaluation refuses to 'globalize' the evaluation of a statement p. We cannot directly say whether a statement is true or false. We will say that *it is determined to be true (or false) at this point in time.*

Naturally, we will admit the persistence of these truths over time: if a statement is determined to be true at time t, it will also be considered true for all instants *subsequent* to t.

But what is time? For what interests us here (the distinction between before and after), it is only a metaphor for an order-structure. This suggests that if p is true at time t, it is true for all times q where q is after t, and therefore where $t \leq q$.

So what then is the 'smallest' time? It is that which is composed solely of two 'instants'; the instant 0 and the instant 1. Here we rediscover our set (0,1) and its standard order-structure, but handled in a completely different way to that of the minimal classical model.

In effect, in the case of the statement p, there will not be two possibilities (true or false) but three:

- p is demonstrably true at the instant 0. By virtue of the rule of persistence, it must also be true in the following instant 1. It is therefore 'constantly' true.
- p is not demonstrably true at the instant 0, but it is demonstrably true at the instant 1.
- p is never demonstrably true.

And this is why, transplanted onto a 'temporal' interpretation of order on (0, 1), our triplet (0, M, 1) comes into play (some values will have the same name as instants, but it won't be the first time we have seen something like this).

If p is constantly true, we attribute to it the value of 1.

If p is only demonstrated at the second instant, namely 1, we will attribute to it the intermediate value M.

If p is never demonstrated, we attribute to it the value 0.

The rules of our minimal model are then perfectly intelligible. Take for example the fact that $-M = 0$, and pose in the usual way that $-M$ interprets the statement not-p when p takes the value M. If p has the value M then it is not demonstrable at the instant 0 but it is demonstrable at the instant 1. But in that case not-p cannot be demonstrated at the instant 0. Since by virtue of its persistence in time, we would also have not-p at the instant 1. Yet we demonstrated p at the instant 1, which violates the principle of non-contradiction.

You may ask: is it true, this principle? Yes, because its substance $LI(x, -x) = 0$ is a demonstrable equation, as we have noted, for the three possible values of x.

For the same reason, not-p cannot be demonstrated at the instant 1, where we have demonstrated p. Thus not-p is never demonstrated, and must have the value 0.

It is therefore entirely consistent that $-M = 0$.

The subjacent rule is clear enough: the negation of p is true at the instant t if at no instant coming *after* t the statement p proves to be indemonstrable.

This also explains rather nicely why excluded middle is invalid. Because if at the instant 0 you can demonstrate neither p nor not-p, but at time 1 you do demonstrate p (or indeed not-p, the result doesn't change), we find that p has the value M and that 'p or not-p' is interpreted as $SS(M, -M) = M$ (and not 1).

This evaluation reflects the fact that at the instant 0, *neither p nor not-p have been demonstrated*, and therefore that 'p or not-p' is not universally valid, since there is a time (0) where this statement is 'in suspense', where neither of the two terms of the disjunction have been demonstrated (or refuted).

Our third value M is also charged with evaluating the temporalization of the true: it affects every non-axiomatic statement (true from time 0), and demands, to be established and thus held as true, at least 'one time'.

Regarding double negation, we know that the value M precludes its being identical to affirmation.

This is illuminated by a profound observation: a statement which takes the value M *is not in fact the negation of any statement evaluable in the (brief) time allotted us*, the time of two instants. See the table, which defines the complements: a complement can only take the values 0 or 1, and never the value M.

In this kind of logic, *negation is not on the same level as affirmation.* Because the evaluation of not-*p* is always temporally dependent on the evaluation of *p*: to know if not-*p* is true or false at time 0, *we must have already ascertained the status of p at times 0 and 1.*

In fact, not-*p* will be valid at time 0 (and therefore at time 1 as well) only if *p* is valid at neither time 0 nor time 1. The evaluation of not-*p* is therefore *retroactive*, with respect to that of *p*.

One consequence of this is that a negative statement *cannot take the intermediate value*: because if, arriving at time 1, we validate *p*, not-*p will have always been false.* And if, arriving at time 1, *p* was not valid, not-*p will have always been true.* Not-*p* can only take the minimal value (0) or the maximum (1).

This 'weakened' conception of negation is typical of intuitionism, for whom the effective *construction* of things is the law of all thought. Negation is only a retroactive sub-product of one such construction (the case where *p* is demonstrable at any instant whatsoever, and where as a result not-*p* will have always been false), or alternatively the notation of the impossibility, *temporally investigated at every instant*, of such a construction (the case where *p* is not demonstrated at any time, and where not-*p* will have always been true).

So the introduction of a third value M only effects statements that do not begin by a negation, the 'positive' or constructive statements, susceptible to being temporally and laboriously verified or constructed.

For intuitionism, a double negation is firstly a negation; its iteration (or doubling) is second. As it is a negation, it therefore takes the values 0 or 1, but never the value M.

We will also say that for intuitionism, the 'negative' dimension of the negation of the negation outweighs its 'double' aspect. It remains dependent and retroactive.

Negation is in the future anterior, which, paradoxically, is positioned, in the doctrine of truth that I proposed in *Being and Event*, precisely on the side of truth.

It remains to project our minimal model into a Topos, to give it the status of a logical matrix of a possible universe.

There are three values (0, 1, M), so there must therefore be three distinct elements of the Central Object, which we will call the true, the false, and the in-extremis (if p is demonstrated at time 1, this is only barely so, since after there is no more time).

Through the bi-univocal correlation J between the elements of C and the sub-objects of 1, there are also three distinct sub-objects (represented by dis-similar monomorphisms).

Note. The astute reader who has followed us through out 'tinkering' in §17 is doubtless wondering: what about the arrow $C \to 1$, which necessarily exists? Is it not a sub-object of 1? And in which case, since C can be isomorphic to neither 1 nor 0, meaning the above arrow $1(C)$ is similar to neither $1(0)$ (which gives the false by centration) nor $\mathbf{Id}(1)$ (which gives the true), *how can there be any bivalent Topoi?* Is not every Topos necessarily trivalent, like the ones we have been trying to construct with the sub-objects $1(0)$, $1(1)$ and $1(C)$?

But we have shown that every well-pointed Topos is bivalent! So are there then no well-pointed Topoi? Yet another logical dilemma.

Naturally, there is a reasonable solution: admittedly, a (unique) arrow exists from C to 1, *but this arrow is not a monomorphism*, and therefore does not define any sub-object.

This is easy to show, since we know that C has at least two distinct elements, T and F. We then have the diagram:

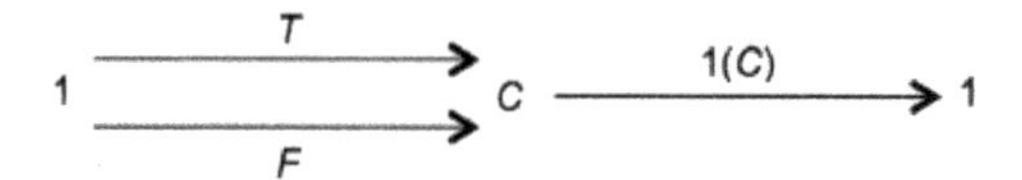

The compositions $1(C) \circ F$ and $1(C) \circ T$, being the arrows from 1 to 1, are equal. If $1(C)$ was a monomorphism that would make $F = T$, and the Topos would be degenerate.

Guided by this minimal model we can tinker a little with our trivalent Topos.

Our third element of C is in extremis, which we will note *inex*. Suppose that our third sub-object is represented by the monomorphism $M \dashrightarrow 1$, which we won't hesitate to call M, as is customary.

We want to have $-M = 0$. This point is crucial for the non-classicism of the Topos, since for 1 and 0 taken 'together', everything operates according to the general necessity of Topos, which also subsumes classicism.

This means that $c(-M) = c(0)$, and as a consequence we have:

centration of $-M$ = the false.

This correlation between the 'intermediate' value and the false is crucial, as it clarifies a point concerning negation (the meaning, in a Topos, of its possible 'weak' value).

For we know that $c(-M) = \backsim \circ c(M)$. This in fact the definition of the complement. It follows that: $\backsim \circ c(M) = F$. However we know that $\backsim \circ T = F$. Therefore we might tentatively say that $c(M) = T$! This would be absurd, since two sub-objects of the same centration are 'the same', and T is already the centration of **Id**(1). In fact, the result would be that 1(M) and 1(1) name the same sub-object and that T and *inex* are the same thing, which would totally eliminate the sought-after trivalence and return us to the classical case.

It is therefore out of the question that $c(M)$ has the value of T. Nor, moreover, can it have the value of F, because 1(M) would then be the same as 1(0). It *must* be that $c(M)$ has the value of the third element of C, namely, in extremis.

We have both equations:

1 $\backsim \circ inex = F$

2 $\backsim \circ T = F$

The generic equation $\backsim \circ x = F$ therefore admits *two* distinct solutions; since *inex* and T are in no way identical.

That the negation of the true has the value of the false doesn't mean in turn that the true is the only one which, when negated, has the value of the false. There is therefore in this case a sort of ambiguity to negation, which gives the same evaluative action (the false) when applied to quite distinct evaluative actions.

In other words, negation makes different evaluations equivocal.

Note. This proves in passing that negation *is not a monomorphism*, and hence not the name of a sub-object of the Central Object. Because the two equations above tell us that $\backsim \circ inex = \backsim \circ T$, which, if the negation was monic, would force $inex = T$, and therefore the disappearance of the evaluative autonomy of *inex*.

In light of the above equation let us say again: two arrows can be distinct, even if their negations are identical.

This establishes once again a dissymmetry between negative equivocity and affirmative univocity, and ultimately between the true and the false. Because the identity of the Central Object **Id**(C), which is the centration of the true, is indeed itself a monomorphism (or even an isomorphism) from C to C. Which, as we have seen, cannot in general be the negation, which is the centration of the false.

Finally, to project our minimal model onto a Topos entirely, it is necessary that the diagram of its sub-objects of 1 present itself as such:

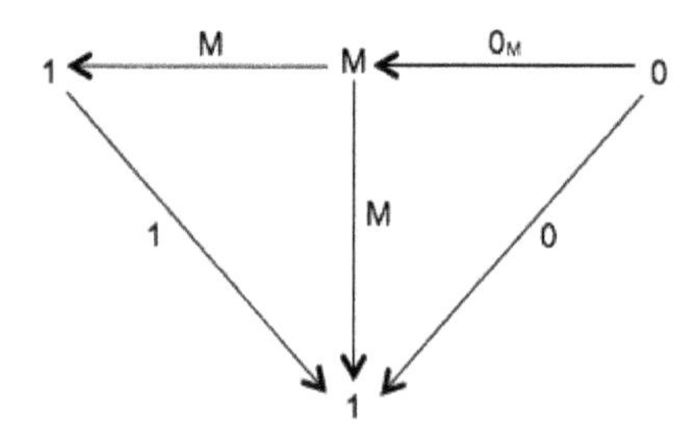

Here we can clearly see (since the triangle obviously commutes) that we have:

$$0 \subseteq M \subseteq 1$$

With respect to the GI of a sub-object and its complement, which must be equal to zero, the case of 1 and of 0 are as normal.

That $M \cap -M = 0$ can be inferred from the fact that, $-M$ being equal to zero, the said intersection (or GI) is the diagonal of the pullback of the arrows 1(M) and 1(0). But this pullback imposes, as the left side of the square, an arrow from the object $M \cap - M$ to 0:

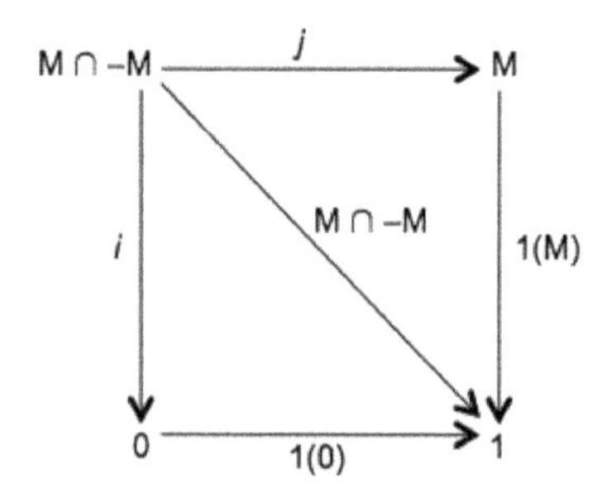

This, as we have known since §9, entails that in every Topos the said object is isomorphic to zero.

Let's return to the crucial point of the 'non-classicism': that the union of M and its complement –M (thus their SS) is not equal to 1. This entails that in such a Topos, the principle of excluded middle is not universally valid.

We can establish this point quickly enough, but it is doubtless useful to take the opportunity to retrace the constitutive elements of union, which involve concepts we have only barely used, like the co-product and the epic-monic decomposition of an arrow in a Topos.

Let's begin with a proposition concerning the co-product in a Topos.

Proposition: the source object of a co-product of the kind $a + 0$ is isomorphic to a. Or in short: $a + 0 = a$.

> *Note*. We could simply say: this proposition is the dual (cf. §6) of the proposition $a \times 1 = a$, because the co-product is the dual of product, and an initial object 0 is the dual of a terminal object 1.
>
> Yet we have already noted (§17) that $a \times 1 = a$. Therefore, by simple dualization, or by passing through the category C^{op}, we equally demonstrate that $a + 0 = a$.
>
> We will however, for practice, demonstrate this *completely*, which we haven't really done for $a \times 1 = a$.

Consider the following commutative diagram which defines the co-limit position of $a + 0$ with regard to a (these are the arrows going from a to a, namely **Id**(a), and from 0 to a, the mandatory and unique arrow 0_a):

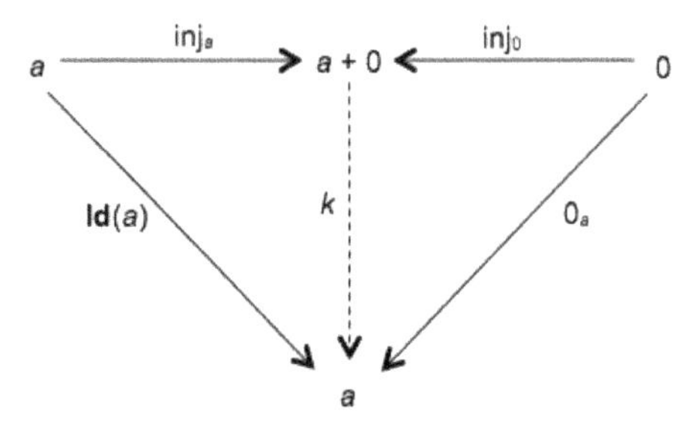

The unique arrow k expresses the co-limit position of $a + 0$ (this is the co-product arrow [**Id**(a), O_a]).

Through commutation we get:

1 $k \circ \mathbf{inj}_a = \mathbf{Id}(a)$

Furthermore, $\text{inj}_a \circ k$ is an arrow from $a + 0$ to $a + 0$. Consider now the co-limit position of $a + 0$ in relation to itself (there are the arrows from a to $a + 0$, namely inj_a, and of course from 0 to $a + 0$), by bringing in the arrow $\text{inj}_a \circ k$:

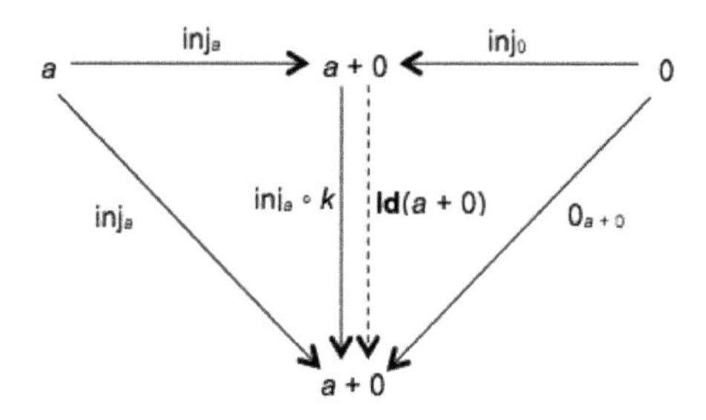

Note that the right triangle definitely commutes (since $\text{inj}_a \circ k \circ \text{inj}_0$ is an arrow from 0 to $a + 0$, though there only exists one single arrow of this kind, which is precisely 0_{a+0}, the third side of the triangle).

But the left triangle also commutes, because:

2 $(\text{inj}_a \circ k) \circ \text{inj}_a = \text{inj}_a \circ (k \circ \text{inj}_a)$ (associative)

3 $(\text{inj}_a \circ k) \circ \text{inj}_a = \text{inj}_a \circ (\mathbf{Id}(a))$ (by 1.)

4 $(\text{inj}_a \circ k) \circ \text{inj}_a = \text{inj}_a$ (identity)

And the triangle commutes.

But equally, the identity $\mathbf{Id}(a + 0)$, placed in the central position, evidently commutes the two triangles.

Yet there must only be one single arrow which carries out these commutations, since $a + 0$ is a co-limit.

Therefore:

5 $\text{inj}_a \circ k = \mathbf{Id}(a + 0)$.

By comparing 1. and 5. above, we see that k, admitting inj_a as its inverse, is an isomorphism, and that a is thereby isomorphic to $(a + 0)$.

The question concerning the union of M and –M (thus of M and 0) can then be resolved without too much difficulty.

Here we must remember that union is the monic part of the epic-monic decomposition of the co-product arrow [1(M), 1(0)]. This is an arrow from M + 0 to 1, which can also be properly presented as the arrow from M to 1, namely 1(M), since M + 0 is isomorphic to M.

We therefore have the diagram:

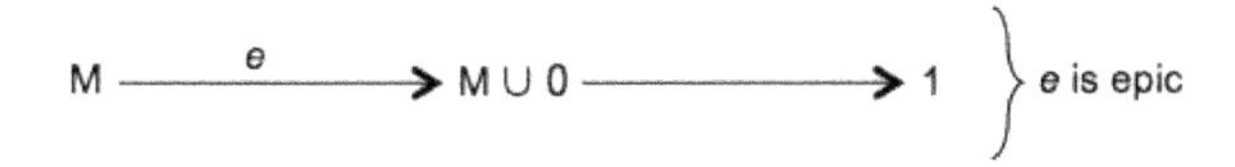

If we now suppose that the sub-object (M ∪ 0) is equal to **Id**(1), examination of the diagram below gives 1(M) = *e*. It follows that 1(M) *is epic.*

But consider the pullback which defines the centration of 1(M), which is *inex*:

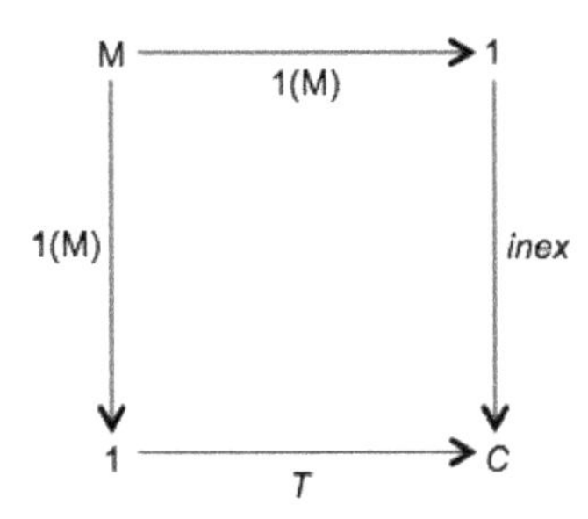

We have (by commutation) $T \circ 1(M) = inex \circ 1(M)$. If 1(M) is epic, it follows that $T = inex$, which completely ruins our construction.

We cannot legitimately suppose that M ∪ 0 = 0, since we must have (by the properties of U): M ⊆ (M ∪ 0); which would not be the case if (M ∪ 0) was equal to 0.

We have shown that (M ∪ 0) cannot be equal to 1 either.

That leaves M as the only possibility.

An examination (which I leave to the reader) of the effects of this conclusion (namely M ∪ –M = M) shows that it is entirely in keeping with everything that we have attempted to cover.

The 'logical kernel' of the trivalent Topos thereby emerges, which, by not validating the excluded middle, is not classical.

It is therefore perfectly true that the concept of Topos proposes possible universes whose immanent logics can be very different.

The concept of Topos subsumes the idea of possible universes without prescribing a determining logic. Its strength lies in placing itself *underneath* the logical options that are organically linked (notably on the basis of the ontology of the void and of difference) to ontological options.

Category theory is a description of the possible options for thought. It does not constitute by itself such an option. In this sense, it is itself a logic, the virtual logic of onto-logical options.

PART TWO

BEING THERE: MATHEMATICS OF THE TRANSCENDENTAL

INTRODUCTION

Let us propose, in the style of the ancient Atomists, that being is pure multiplicity, without ground or meaning. Taking this loss of the One to its ultimate consequences: experience is not constitutive, there is no generic Life. There is only the flat surface of indifferent multiplicity. Ever since the Greeks the thought of this surface carries the name of mathematics: that which concerns itself only with the multiple-being of a multiplicity, to the exclusion of all qualities and intensities.

Under these conditions, to be is, for a multiple, to belong to another multiple whose being is already presupposed. We name 'situation' this referential multiple whose prodigality is such that it gives its bit of being to anything inscribed within it as an element. We will then say that to be is to belong to a situation. Retaining the formalism of set theory as it is taught in schools (whose suitability for an ontology of indifferent multiplicity is not in doubt): if ε is the multiple we are sure it is, and if S is the referential situation, the statement which assures the being of ε is written: $\varepsilon \in S$.

However, this assertion of being does not tell us in what sense the element ε *exists*. If 'to exist' does not mean the same as 'to be', then it is to the pure neutrality of multiple-being (which it obviously requires) that existence adds something. This addition cannot be 'quantitative', because the notion of quantity is entirely subsumed by the multiple composition of any singular being. In truth, there is no ontological distinction between being and quantity (provided, of course, that we suppose – in strict fidelity to Cantor – that the infinite is quantifiable, and that, following the same line of thought, we also know to differentiate quantity from number, which is merely one form of quantity among many others, and not the most important). It follows from all this that existence is nothing if it is not a *quality* of being.

But every quality implies variability and nuance. Every quality is modal. Multiple-being, whose determination is the ontology of quantity as such, ignores modality. Either a multiple, insofar as it belongs to a situation, *is* (relative to that situation), or else it does not belong, and as such *is not*. The precise formula for this exigency is that there is no measure of being which is ontologically expressible. Once we have supposed the concept and distinguished it from being, existence names exactly that which, ontologically, is not: a degree of being. Nor does this degree in any way affect the being of being. It is an index of *its appearance* (in the situation). This is why existence is always a matter of *more or less*. Given a multiple ε, its existence, which is its degree of non-being of being, and therefore the value of appearance of the formula $\varepsilon \in S$, disposes itself between 'absolute' existence (ε's belonging to the situation S appears as certain and complete, the formula $\varepsilon \in S$ is *proven true*) and inexistence (ε's belonging to the situation S appears void, the formula $\varepsilon \in S$ is proven entirely false: the being ε inexists in S). These two extreme polarities of existence in no way rule out there being other more or less numerous and nuanced degrees of existence, which amounts to saying that the multiple ε exists in S, 'to a certain degree'. It all depends on the existential resources of the situation S.

What exactly is the 'existential resource' of the situation S? To understand this, we must once again think the requirement, for every multiple, of only being 'present' – thus being, pure and simple – insofar as it belongs to a situation. This requirement results from the fact that *there is no Whole*. That there is no referential totality, no unique and stable universe (nor for that matter any universal History), is a point that is demonstrable in the mathematical theory of the pure multiple: it is inconsistent to suppose a set of all sets, which means: total multiplicity (which would be, in reality, a One-All) cannot come to be.

But if there is no Whole, the being of a multiplicity, having no 'cosmic' place where it is affirmed, assumes a particular or local referent. Being is necessarily linked to its preventative localization, whose space is the situation to which it belongs. Being is essentially being-there (*Da-sein*). And what fixes the localization is only ever another multiple 'already-there' (whose being is proven), and which is the S of every ε, its situation.

Now, how does the situation localize the multiples that belong to it? In the first (ontological) sense, it is simply a matter of belonging or non-belonging. A particular being is apportioned being insofar as it belongs to a situation, and non-being insofar as it does not belong to it. But this bears only on the being of being-there. 'Globally' this is because the being of beings is derived from the situation. We can now ask: what is the *there* of being-there? How, locally, is it that a being comes to be there? Not only there 'in the situation', which guarantees its being, but as the singularity of being-there in that situation. We can only respond to this question if, in a given situation, the localization of a multiple-being is fixed by a network of relations which determine the *difference* of that being to the other elements of the situation. Yet 'belonging to the situation' says nothing of this difference, if these other beings also derive their being from this belonging.

Let's fix our terminology. We will call *appearing* that which, of a being as such (a mathematical multiple), is seized in a situated relational network, such that we can say that this being is more or less different from another multiple that belongs to the same situation.

Ontologically, one multiple cannot be 'more or less' different to another multiple. A multiple is only identical to itself, and it is a law of being *qua* being (axiom of extensionality) that the least local difference, bearing for example on one single element amongst an infinity of others, entails an absolutely global difference. It will not be the same for appearing. It is clear that, in a situation, multiples can be more or less different, similar, proximate, etc. We must therefore admit that what governs appearing is not the ontological composition of a particular being (a multiple), but the relational evaluations that determine the situation and localize this being within it. Unlike the legislation of the pure multiple, these evaluations do not always equate local and global difference. They are not ontological. This is why we will call logic the law of the network of relations that determines appearing in the situation of multiple-being. Every situation possesses its own logic, which legislates on appearing, or the 'there' of being-there.

The minimum requisite for any localization is we can set a degree of identity (or of non-identity) between an element α and an element β, supposing that both belong to the situation. There are grounds for thinking that there exists, in any situation, what we will call a *function of appearing*, which, given two elements of the situation S, measures their degree of identity. We will write $\mathbf{Id}(\alpha, \beta)$ for this function of appearing. It assigns the measure according to which, and pursuant to the logic of the situation, we can say that the multiples α and β appear identical.

But what are the values of the function of appearing? What *measures* the degree of identity between two appearances of multiplicities? Once again, there can be no general or totalizing answer. The scale of evaluation of appearing, and thus the logic of a situation, depends on the situation. What we can say is that such a scale exists in every situation.

We will call *transcendental* of the situation that which, in any situation, serves as a domain for the evaluation of identities and differences in appearing. Like everything, the transcendental is a multiple, which obviously belongs to the situation of which it is the transcendental. But this multiple is endowed with a structure which authorizes the arrangement, on its basis, of the values (the degrees) of identity between the multiples which belong to the situation, such that we can fix the value of the function of appearing **Id**(α, β), whatever α and β may be.

This structure has properties that vary according to the situation. But it also has general invariant properties, without which it could not operate. There is a general mathematics of the transcendental. The purpose of this book is to present the broad outlines of this mathematics.

Ultimately, the concept of existence and its properties depends on this. For what does it mean for a multiple ε 'to exist'? It means: to appear in a situation as identifiable in it. And thus: being transcendentally evaluated in the situation on the basis of its identity to itself. A multiple exists in a situation S to the extent that its identity to itself takes such and such a value. However, as the identity of a multiple α to a multiple β is measured by the value of the function of appearing **Id**(α, β), so too the degree of identity to itself of a multiple ε is obviously measured by the value **Id**(ε, ε). The definition of existence is therefore precise and formalizable: *we call existence of a given multiple, relative to a situation, the degree according to which, in that situation, the multiple appears as identical to itself.* Formally, the existence of a being ε, relative to a situation S, is the value of the function of appearing **Id**(ε, ε) in the transcendental of the situation. To express this derivation of existence on the basis of the question of identity in appearing – or logical identity of a being – we will write **Id**(ε, ε) in the form **E**ε, and we will say that this is a function of existence of the situation, applied to multiple-being ε which appears there.

It is essential to see that existence is not a category of being as such, but rather a category of appearing. Or, more rigorously, that existence concerns the logic of being, and not its ontological status. It is only according to its being-there, and not according to its multiple composition, that a being can be said to exist. And this is always, at the same time, a degree of existence, situated between inexistence and absolute existence. Existence is at once a logical and an intensive concept.

It is precisely in order to acquire a genuine understanding of this double determination (logical and intensive, or equally: algebraic and topological) that it is critical to develop the mathematics of the transcendental.

The general plan, in five sections, is the following:

A. Structure of the Transcendental (Heyting Algebra)
B. Transcendental Connections
  B1. The Transcendental and the Theory of the Pure Multiple
  B2. The Transcendental and Normal Logic
  B3. The Transcendental and Topology

C. Transcendental Operations. Theory of Appearing and of the Object
D. Transcendental Projections. Theory of Localization
E. Theory of Relation. Situation as Universe
Appendix. Three Concepts of Identity

No special mathematical knowledge is required outside of what we should all have been taught, namely, the ability to follow a formal definition or demonstration.

I note once and for all the essential mathematical material I put to work here for my own philosophical ends is dispersed throughout Robert Goldblatt's book, *Topos: The Categorial Analysis of Logic.*[40] The work of several other mathematicians has aided this elaboration, amongst whom Fourman, Higgs and Scott should be cited.[41]

In the present text, the successive stages of the demonstrations are generally presented as exercises. So, Ex.B.3 names exercise 3 of section B. This exercise is feasible strictly in relation to what has already been established in the text. We recommend the reader attempt the demonstration themself, without turning immediately to the completed demonstration. Thus, the reader will gradually be able to really think from within the structures of the transcendental.

The fundamental kernel of this text, the veritable armature for the philosophical thesis of the being of appearing, resides at the start of section C, which supposes a reasonable familiarity with section A.

A. TRANSCENDENTAL STRUCTURES

A.1. The transcendental T of any situation is, in its being, a multiple. Of course, this multiple belongs to the situation, it is presented there. But beyond this ontological characterization, the transcendental legislates, in the situation, on the degrees of appearing, the nuances of quality or the intensities of existence. We do not yet know exactly what this means. However, we can presume that the transcendental makes possible evaluations and comparisons, that it composes a scale of measure of the more and the less. And we know that the simplest abstract form of such a power is the order relation, which makes it possible to say that the element β is 'greater' (or situated 'higher' in the scale of comparison, or has a superior intensity etc.) than another element δ.

To fully grasp the essence of the order-relation, it is useful to compare it to the other primitive relation, that which sets the strict (or rigid) identity between two elements. We call this the equivalence-relation. It axiomatically rules on the identity (the equivalence) of two elements β and δ in the following way:

- **a** An element β is always identical to itself (reflexivity)
- **b** If β is identical to δ, and δ to π, then β is identical to π (transitivity)
- **c** If β is identical to δ, then δ is identical to β (symmetry)

We will note that the relation of equivalence imposes a rigorous symmetry, formalized in the axiom properly named 'axiom of symmetry'. In this relation, the relation of one element to another is the same as that of the relation of this other to the first. This is why the most frequent usage of the equivalence-relation is to permit the *substitution* of δ by β in a formula, once we know that δ is equivalent to β. We could even say that our three axioms (reflexivity, transitivity and symmetry) are axioms of substitutability.

But it must be concluded that the very essence of relation, the essence of every differentiated evaluation or every comparison, is not yet captured by the 'relation' of equivalence. For a comparative evaluation supposes that we can compare really distinct – that is to say non-substitutable – elements. But this means rejecting the third axiom, which affirms the symmetry of related elements.

The order-relation is essentially the result of this rejection. It is 'like' the equivalence-relation, except that it replaces symmetry by anti-symmetry.

The order-relation assumes that difference is axiomatically perceptible. Of course, that a term is related 'to itself' can constitute a primitive given. Similarly, that the relation 'transits' (in the sense that if $\beta = \delta$ and $\delta = \pi$, then $\beta = \pi$) is a useful property of expansion. We therefore retain reflexivity and transitivity. But ultimately, it is there where two terms cannot be substituted in terms of what links them that the relationship between relation and singularity is affirmed, and that differentiated evaluations become possible. Therefore we explicitly reject symmetry.

A.2. A relation between elements of a set A is an order relation, noted $\leq$, if it obeys the three following axiomatic dispositions:

1. Reflexivity: $x \leq x$
2. Transitivity: $[(x \leq y) \text{ and } (y \leq z)] \rightarrow (x \leq z)$
3. Antisymmetry: $[(x \leq y) \text{ and } (y \leq x)] \rightarrow (x = y)$

Antisymmetry is what distinguishes an order-relation from equivalence, and marks our real entry into the domain of the relation between non-substitutable singularities. In the (order) relation that x maintains with y, the element x can change its place with y only if the two elements are rigorously 'the same'. The order-relation is in fact the very first inscription of an exigency of the Other, insofar as the places that it determines (before $\leq$ or after $\leq$) are in general not exchangeable.

Reflexivity and transitivity are Cartesian dispositions: self-relation and chain of reasons. But these are properties that also apply to identity or equivalence. The order-relation retains these Cartesian properties. But with antisymmetry it formalizes a certain type of non-substitutability.

For intuitive convenience, and in order to see the comparisons, $x \leq y$ can be read in one of the two following ways: 'x is less than or equal to y', 'y is greater than or equal to x'. We could even say, 'smaller than' or 'greater than'. However we should take care that the dialectic of the great and the small does not come to subsume the entire, axiomatically determined field of the order relation. These are only ways of speaking (orally). The essence of the order-relation is comparison 'in-itself'.

It is not necessary that an order-relation binds together all the elements of the base set A. If this is the case, we say that the relation is total. In the case of a total relation in which x and y are two random elements of A, we always have either $x \leq y$, or $y \leq x$.

If the relation is not total, we say that it has a partial order, and A, endowed with such an order-relation, is called *a partially ordered set*, or POS.

In a partially ordered set (POS), we have three possible cases for any two elements x and y: either $x \leq y$, or $y \leq x$, or neither $x \leq y$, nor $y \leq x$. In the third case, we say that x and y are incomparable (or unrelated).

We posit that *the transcendental T of any situation is a partially ordered set (POS)*. This is the basic structure of any multiple functioning as a transcendental.

A.3. Maintaining the supposition that, with regard to what appears in the situation, the transcendental T supports evaluations of intensity, it is reasonable to assume the capacity to determine a *nil intensity*. Of course, nil or zero are determinations relative to the situation, and therefore to the transcendental of that situation. In the order of appearing (or of logic), a nil intensity 'in itself' makes no sense. We know that there is in the order of being an intrinsic emptiness, the empty set, or set of nothing, which is 'in itself' the minimum multiple. But in reality the empty set is a name. And this name only makes sense in *one* very particular situation, the ontological situation, which is the historical development of mathematics. In this situation, 'empty set' or $\emptyset$ is the proper name of being *qua* being. With regard to what appears in any situation, all that we can hope for is the power to transcendentally evaluate a *minimal* intensity. Obviously, for the 'inhabitants' of this situation, such a minimum equates to zero, since they will have no conception of any intensity less than the minimum. But for the ontologist (the mathematician), this minimum will be relative to the transcendental under consideration.

We therefore propose that the transcendental T of any situation composes a minimum for the order which structures T. Let μ be this minimum. It marks an element such that, for any element p of T, we have $\mu \leq p$. (μ is 'smaller' than any element p of T.)

Ex.A.1. Show that if the minimum μ exists, it is unique.

Let's suppose that both μ and μ' are minimum. Then by the definition of the minimum, $\mu \leq \mu'$ and $\mu' \leq \mu$. The axiom of anti-symmetry requires that μ and μ' be the same, thus $\mu = \mu'$.

A.4. If T must provide comparisons, it must be able to indicate that which is 'immediately smaller' than any two given elements. The underlying phenomenal idea is to express what is common to two elements appearing in a situation. Or that which is, by its degree of intensity, or by qualitative degree, enveloped by the degrees of appearing of these two elements.

This idea establishes that for any POS (here, for a transcendental T): given two elements p and q of T, we suppose that there always exists – and we write this $p \cap q$ – an element which is the largest of all those which are inferior to both p and q. In other words, if $r \leq p$ and $r \leq q$ (r is inferior to p and q), we have, on the one hand, $p \cap q \leq p$ and $p \cap q \leq q$ ($p \cap q$ is itself also inferior to p and q), and, on other hand, $r \leq p \cap q$ (which means that $p \cap q$ is superior to any r which has that property). We will also say that $p \cap q$ is the GI (greatest inferior) of p and q.

For a given p and q, that their GI is unique, as for the minimum μ, is an immediate consequence of the axiom of anti-symmetry.

We see that, in the Transcendental T, the GI (namely $p \cap q$) measures the degree of appearing 'immediately inferior' to the degrees p and q.

We will note that in the definition of $p \cap q$, p and q play exactly the same role. This definition is symmetrical. Consequently, we have $p \cap q = q \cap p$. We will say that the operation GI is commutative. It is moreover trivial that $p \cap q = p$, since $p \leq p$. The GI operation is *idempotent*.

Ex.A.2. Show that $p \cap q = p$ is equivalent to $p \leq q$. (If the GI of p and of q is p, then p is smaller than q, and if p is smaller than q then $p \cap q = p$.)

- Direct proposition: if $p \cap q = p$, then $p \leq q$
  By definition of $p \cap q$, we have $p \cap q \leq q$. Thus, if $p \cap q = p$, we get $p \leq q$.
- Reciprocal proposition: if $p \leq q$, then $p \cap q = p$

We know that $p \leq p$. We can then say that p is the largest element which is lesser than or equal to p. If on the other hand $p \leq q$, it is certain that p is the largest element simultaneously lesser than or equal to p and q. This is the definition of $p \cap q$.

A.5. After the *analytic* determination of the elements of p and q, expressed by the GI $p \cap q$ (degree of intensity situated just beneath the degrees p and q), we have need of a more synthetic determination in order to construct logic as the legislation of appearing. The intuitive phenomenal idea is to express, through a single intensity, *the entire intensity contained within a fragment or part of the situation*, but also to express it 'as precisely as possible', to grasp as closely as possible this sum of intensities. This involves finding the intensity of appearing of an element that *envelopes* the global appearance of the part concerned. Or, put another way, fixing the degree of appearing immediately superior to that of each element contained in this part.

We thus attain this intuition: Let us consider the subset B of the transcendental *which represents all the intensities of appearance of the fragment of the situation under consideration*. Or again, if $S_1 \subseteq S$ is a (finite or infinite) part of the situation, $B \subseteq T$ will be the part of the transcendental which contains all the measures of intensity of appearing of the elements of the part S_1. We will then directly look for a measure of intensity that 'envelops' all the intensities contained in B.

Let us suppose that there exists at least one element t of T which is superior to (greater than or equal to) all the elements of B. In other words, if $b \in B$, then $b \leq t$. We will say that t is an *upper bound* of B.

Suppose now that there exists an element u which is the *smallest of all the upper bounds of* B. In other words, u is an upper bound of b, and if t is another upper bound of B, we have $u \leq t$. We say that the element u is the *envelope* of B. We also say that B is a *territory* for u. By an abuse of language we can even say that a situational fragment S_1, whose intensities of appearing are summed up by B, is enveloped by the intensity u, or by any element whose intensity of appearing is u. In other words, the intensity u has the part S_1 of the situation as its territory.

If the envelope u exists for a subset B of the transcendental T, then, by the axiom of symmetry, it is unique.

We then propose that *in a transcendental T, any subset B of T admits an envelope u. Or there always exists a u for which B is a territory.*

We can envisage this property as one of *phenomenal completeness*. We are able to think (to measure) the envelope (the intensive synthesis) of any phenomenal presentation, whatever its ontological characteristics.

Ex.A.3. Show that the existence of an envelope for any part of T leads to the existence of a maximum of T, and therefore of the measure of a maximal intensity of appearing in the situation S of which T is the transcendental.

It is an ontological principle that any set can be considered as a subset of itself. Consequently, we have $T \subseteq T$. There must therefore exist an envelope of T, thus an element of T, nominally M, which is an upper bound of T. This means that M is superior or equal to any element of T. Thus, if $p \in T$, then $p \leq M$. This element M is therefore a maximum in T. That it is unique is demonstrated (as with the minimum μ, cf. Ex.A.1.) by the axiom of symmetry.

In general we write $u = \sum B$ to indicate that u is the envelope of B, or that B is a territory for u.

B can be defined by a certain property of elements of T. For example, B can be 'all the elements of T which have the property P'. We will then write $B = \{q \in T \,/\, P(q)\}$, to be read '$B$ is the set of elements q of T which possess the property P'.

In this case, the envelope of B is written: $u = \sum\{q \,/\, P(q)\}$.

We can clearly see that u is the element of T *that is*:

- Greater than or equal to any element q of T, which has the property P (upper bound).
- Less that or equal to any element t which is, like itself, greater than or equal to any element q having the property P (least upper bound).

A specific case is where B is reduced to two elements, p and q. An upper bound is then simply an element t superior to both p and q. And the envelope is the least upper bound, that is the smallest of all the elements superior or equal to both p and q. We call this the SS (smallest superior) of p and of q, and in general we write this $p \cup q$.

Since, in a transcendental, the envelope exists for any subset B of T, obviously the SS of p and of q always exists.

Ex.A.4. Write the SS of p and q in the set-theoretic terms indicated above.

If $B = \{p, q\}$, we can say that the characteristic property of any element of B is to be equal to either p or q. This property is written: '$x = p$ or $x = q$'. The envelope of B can then be written as: $u = \sum\{x \mid x = p \text{ or } x = q\}$

A.6. A remarkable consequence of the existence, for every subset B of T, of the envelope $\sum B$ is the possibility of defining, in the transcendental, an operator which, when we identify the logical signification (in the classical sense) of the transcendental, we will use to interpret implication, which is the fundamental connective of every deduction.

First let us give the formal procedure. Let p and q be two elements of T. Let us consider the subset B of T so defined: 'all the elements t of T whose GI with p is inferior to q'. This is written: $B = \{t \mid p \cap t \leq q\}$. We will note that B is never empty, since $p \cap q \leq q$, meaning B consequently contains as elements at least p, $p \cap q$, and, of course, the minimum μ.

As every envelope in a transcendental exists, the envelope of B must also exist. We call this the *dependence* of q with regard to p, and we write this $p \Rightarrow q$.

The underlying idea is the following: the envelope of the above B is (approximately!) the largest element t such that $p \cap t \leq q$. This is in fact the largest 'piece' that we can connect to p while remaining 'close' to q. We therefore have here a measure of the degree of dependence of q with regard to p, or the degree of possible causal proximity between p and q. Or to put it another way, the largest degree of intensity which, when combined with the one measured by p, remains inferior to the intensity measured by q.

It is important to note that, once p and q are fixed, the dependence $p \Rightarrow q$ is itself a fixed term of T, and not a relation between p and q. This means that $p \Rightarrow q$ is the fixed result of the envelope-operation given by $\sum\{t \mid p \cap t \leq q\}$, an operation which concerns p and q.

Ex.A.5. Show that if $p \leq q$, then the dependence of q with regard to p has the maximum value in T. This is written: $p \Rightarrow q = M$.

Regardless of what t is, we have $p \cap t \leq p$ (definition of $\cap$). If now $p \leq q$, whatever t may be, we have $p \cap t \leq q$ (transitivity). It follows that the subset $B = \{t \mid p \cap t \leq q\}$ is equal to T as a whole (since every $t \in T$ has the property which defines the elements of B). Its envelope is therefore the maximum M, which has been precisely defined as the envelope of T (Cf. A.5). But the envelope of B is, by definition, the dependence of q with regard to p. It is therefore established that if $p \leq q$, then $p \Rightarrow q = M$.

Remark: A rational consequence that we draw from this exercise – especially if we think of the logical interpretation of dependence (material implication) – is that the dependence of p with regard to p itself is the maximum. In effect, we always have $p \leq p$ (reflexivity). Thus $(p \Rightarrow p) = M$.

We now have at our disposal the following structure of the transcendental:

- A minimum μ for evaluating what absolutely inappears in a situation, and a maximum M for that which is unquestionably there.
- A local element common to two apparent elements ($\cap$).
- A global envelope for every apparent (even infinite) multiplicity marked $\sum$.

The relation between the local (or finite) operator $\cap$ and the global envelope $\sum$ is one of distributivity. What an element and an envelope have in common is the envelope of what this element and all the elements that the envelope envelops have in common. This is written: $p \cap \sum B = \sum\{(p \cap b) \,/\, b \in B\}$.

In plain language: the GI of an element p and of the envelope of a subset B is equal to the envelope of the subset of T comprising all the GI's of p with all the elements of B.

This relation is axiomatic: it is not deducible from the intrinsic properties of $\cap$ and of $\sum$.

If the envelope is a local SS (cf. A.6), noted $q \cup r$, distributivity is written as follows: $p \cap (q \cup r) = (p \cap q) \cup (p \cap r)$.

Ex.A.6. (Reciprocal of Ex.A.5.) Show that if the dependence of q in relation to p is equal to the maximum, then p is less than or equal to q. In other words: $(p \Rightarrow q = M) \rightarrow p \leq q$.

If $p \Rightarrow q = M$, any given $t \in T$ is less than or equal to it, since M is the maximum of T. In particular, $p \leq p \Rightarrow q$. By virtue of exercise A.2., this can also be written $p \cap (p \Rightarrow q) = p$. But the definition of dependence of q with regard to p gives:

$$p \Rightarrow q = \sum\{t \,/\, p \cap t \leq q\}$$

Thus, by combining these two remarks, we finally have:

$$p \cap \sum\{t \,/\, p \cap t \leq q\} = p$$

Applying the axiom of distributivity, this becomes:

$$\sum\{p \cap t \,/\, p \cap t \leq q\} = p$$

The element q is obviously an upper bound for all the $p \cap t$ such that $p \cap t \leq q$. If p is the envelope of the subset constituted by these $p \cap t$, it is by definition the least upper bound, and we therefore have $p \leq q$.

Ex.A.7. Show that we always have $p \cap (p \Rightarrow q) \leq q$.

$p \Rightarrow q = \sum\{t \,/\, p \cap t \leq q\}$	Def. of $\Rightarrow$
$p \cap (p \Rightarrow q) = p \cap \sum\{t \,/\, p \cap t \leq q\}$	Application
$p \cap (p \Rightarrow q) = \sum\{(p \cap t) \,/\, p \cap t \leq q\}$	Distributivity
$p \cap (p \Rightarrow q) \leq q$	

Ex.A.8. Show that we have always $q \cap (p \Rightarrow q) = q$.

$q \cap (p \Rightarrow q) = \sum\{q \cap t \,/\, p \cap t \leq q\}$	Def. and distributivity

Let's note that q belongs to the subset of elements of T whose envelope is under consideration here. We can in fact present this in the form $q \cap q$. On the other hand, this satisfies the condition $p \cap t \leq q$, since the definition of $\cap$ implies $p \cap q \leq q$. But q is also the largest of these elements, since for any t we have $q \cap t \leq q$. It is then obvious that q is the envelope of the subset, since it is the (internal) maximum. This allows us to conclude: $q \cap (p \Rightarrow q) = q$.

A.7. Once we have at our disposal the element 'dependence of q with regard to p', thus $p \Rightarrow q$, we can define what we will call the *reverse* of an element p of the transcendental, and we write this $\neg p$. This element will later be used to interpret negation, in its traditional logical sense.

The underlying intuition is that the reverse of an apparent element is the largest element which, in appearing, is totally disjoint from this first element. This forms something of the background upon which the initial phenomenal element appears in the situation: at once linked to its appearance, and totally separated from this appearance as such.

Technically, the procedure is the following: we call the reverse of p, noted $\neg p$, *the dependence of the minimum with regard to* p, thus the element $p \Rightarrow \mu$.

To understand why this is, we must return to the definition of dependence. The reverse of p is written:

$$\neg p = p \Rightarrow \mu = \sum\{q \mid p \cap q \leq \mu\}$$

But as μ is the minimum, this means:

$$\neg p = \sum\{q \mid p \cap q = \mu\}$$

$\neg p$ is therefore the smallest of the elements of T which are greater than or equal to all the q's whose GI with p is equal to the minimum. Or, metaphorically, the reverse of p is the largest of the elements of the transcendental T having 'nothing in common' with p. This is indeed why the reverse serves to evaluate what, in appearing, is given in a situation as the negation of whatever element's intensity is evaluated by p. Because the reverse, as we have defined it, combines the 'nothing in common' with maximality. This is, basically, maximal alterity.

Let's demonstrate some properties of the reverse of an element.

Ex.A.9. The reverse of the maximum is the minimum, thus $\neg M = \mu$.

$\neg M = M \Rightarrow \mu$	Def. of $\neg$
$\neg M = \sum\{q \mid M \cap q \leq \mu\}$	Def. of $\Rightarrow$
$\neg M = \sum\{q \mid M \cap q = \mu\}$	Def. of μ
$\neg M = \sum\{q \mid q = \mu\}$	$M \cap q = q$
$\neg M = \sum\{\mu\} = \mu$	

For the concluding equation we will note that $\sum\{p\} = p$, because $p \leq p$.

Ex.A.10. The reverse of the minimum is the maximum, thus $\neg\mu = M$.

$\neg\mu = \mu \Rightarrow \mu$ Def. of $\neg$

But we have seen above (remark in exercise A.5) that, for any p, $p \Rightarrow p = M$. Thus $\neg\mu = M$.

Ex.A.11. The GI of p and of $\neg p$ is equal to the minimum, that is, $p \cap \neg p = \mu$.

$p \cap \neg p = p \cap (p \Rightarrow \mu)$	Def. of $\neg p$
$p \cap (p \Rightarrow \mu) \leq \mu$	Ex.A.7 $p \cap (p \Rightarrow q) \leq q$
$p \cap \neg p = \mu$	Consequence

Ex.A.12. An element is always lesser than or equal to the reverse of its reverse, that is: $p \leq \neg\neg p$.

$\neg p = p \Rightarrow \mu$	Def. of $\neg$
$\neg\neg p = \neg p \Rightarrow \mu$	Application
$\neg\neg p = \sum\{q \mid q \cap \neg p = \mu\}$	Def. of $\Rightarrow$

But we have seen (in the preceding exercise) that $p \cap \neg p = \mu$. Therefore, p is a part of q such that $q \cap \neg p = \mu$. And as $\neg\neg p$ will be greater than or equal to any q (definition of $\sum$), the result is that $p \leq \neg\neg p$.

VERY IMPORTANT: In general (in a given transcendental), there is not $\neg\neg p \leq p$. Therefore it is not true, most of the time, that the reverse of the reverse of p is less than or equal to p. This is not deducible from the fundamental structures of the transcendental. If we had $\neg\neg p \leq p$, since we have $p \leq \neg\neg p$, anti-symmetry would mean that $\neg\neg p = p$. We would therefore have the 'classic' case where the reverse of the reverse is none other than the initial element. (In the logical interpretation of section B.2, we will say that the logic of appearing is not necessarily classical. Or that the negation of the negation does not always equal affirmation.)

TECHNICAL NOTE: We will come to see that the structure of the transcendental has the following principles:

1. T is a partially ordered set ($\leq$ is not necessarily a total relation).
2. There exists, for this order, a minimum μ and a maximum M.
3. The GI of any pair of elements of T, noted $p \cap q$, exists in T.
4. For every subset $B \subseteq T$, there exists an envelope $u = \sum B$.
5. Distributivity: $p \cap \sum B = \sum\{(p \cap b) \mid b \in B\}$

Mathematicians call this structure a 'complete Heyting algebra'. It serves as the interpretive model for the formalizations (invented by Heyting) of intuitionist logic.

B. TRANSCENDENTAL CONNECTIONS

B.1. Connections between the transcendental and set-theoretic ontology: Boolean algebras

B.1.1. The purpose of this section is first of all to show that a transcendental is defined by certain canonical operations on the *parts* of a set E. We will then ask *what kind* of transcendental allows itself to be directly represented by the multiple form of being. We will see that this is a very particular transcendental, in fact that same one Boole identified when he sought to express classical propositional logic algebraically. Finally, we will establish that *every* 'Boolean' transcendental is expressible in a set-theoretic form, and that, in this sense, there is a kind of legislation of appearing which is isomorphic to that of its being. A Boolean transcendental expresses what we might call the logical type of ontology.

B.1.2. Take any given set E. Let $P(E)$ be its set of subsets (or parts). Or rather, the set of sets X, such that $X \subseteq E$.

The relation of inclusion $\subseteq$ defines an order on $P(E)$. Remember that $X \subseteq Y$ means: every element of X is an element of Y. In effect:

$X \subseteq X$	(reflexivity)
$(X \subseteq Y)\ \&\ (Y \subseteq Z) \rightarrow X \subseteq Z$	(transitivity)
$(X \subseteq Y)\ \&\ (Y \subseteq X) \rightarrow X = Y$	(antisymmetry)

Antisymmetry is perhaps the least obvious. In fact, $X \subseteq Y$ means: every element of X is an element of Y. But $Y \subseteq X$ means: every element of Y is an element of X. If we have both, then X and Y have the same elements. But then (extensional conception of the identity of multiples) they are identical.

The set of the parts of E is therefore a POS. We will show that this POS is, axiomatically, a transcendental (a Heyting algebra).

a Minimum: this is the void or empty set $\varnothing$, since for any set X, whatever it may be, $\varnothing \subseteq X$.

This is a subtle point. Due to the fact that $\varnothing$ has no elements, we can always assert that its elements (hence, 'nothing') are amongst the elements of X. In the absence of anything that could prevent it, there is an inclusive omnipresence of the void.

b GI of two elements. This is the intersection $X \cap Y$, that is, the set of all the elements which are common to X and Y. It is in fact clear that if $A \subseteq X$ and $A \subseteq Y$ then the elements of A are common to X and Y. Thus $A \subseteq X \cap Y$, since $X \cap Y$ comprises all the elements that have this property. So every element of A without exception is an element of $X \cap Y$. It follows that $X \cap Y$ is the largest set inferior to both X and Y. It is indeed the GI (greatest inferior) of X and Y.

c Envelope of a subset of parts. Let I be a (potentially infinite) set, which we will call *set of indices*. A family of subsets of E will be noted $X_{i \in I}$, each of these subsets being assigned an element of I that 'names' it. This is commonly done when we enumerate, for example, the objects $O_1, O_2, O_3, \ldots$ The indices are then whole numbers, and I is the set N of these numbers. All we have done here is generalize this procedure. It is therefore possible to consider their reunion $\mathbf{U}X_i$, namely the set of all the elements that belong to at least one of the members X_i of the family. Formally, an element of the union belongs to an X of the family, which is indexed by an $i \in I$. Therefore, using the sign $\exists$ for 'there exists':

$$x \in \mathbf{U}X_i \leftrightarrow (\exists i)\ [x \in X_i]$$

That $\mathbf{U}X_i$ is an envelope for the subsets X_i is obvious.

On the one hand, for every i, $X_i \subseteq \mathbf{U}X_i$. In fact, $x \in X_i \leftrightarrow x \in \mathbf{U}X_i$ (see above). Therefore every element of X_i is an element of $\mathbf{U}X_i$.

On the other hand, take A such that $X_i \subseteq A$ for every i. This means that if $x \in X_i$, then $x \in A$, and the same goes for every i. Thus $\mathbf{U}X_i \subseteq A$.

Finally, $\mathbf{U}X_i$ is larger than every X_i, and also the smallest of those larger than every X_i. It is the envelope of the family $X_i \in I$. Or, the family $X_{i \in I}$ is a territory for $\mathbf{U}X_i$.

d Distributivity of conjunction relative to the envelope. This is a matter of showing that: $Y \cap \mathbf{U}X_i = \mathbf{U}(X_i \cap Y)$.

This is intuitive, since $Y \cap \mathbf{U}X_i$ is the set of elements which belong to both Y and $\mathbf{U}X_i$. But belonging to $\mathbf{U}X_i$ means: belonging to at least one X_i. Thus, if $x \in Y \cap \mathbf{U}X_i$, this means that $(\exists i)$ $[x \in X_i$ and $x \in Y]$. Therefore $(\exists i)$ $[x \in (X_i \cap Y)]$. This is (cf. above) exactly equivalent to $x \in \mathbf{U}(X_i \cap Y)$.

We can finally see that the POS defined on $P(E)$ – the set of parts of E, ordered by inclusion $\subseteq$ – does indeed have the structure of a transcendental. We derive in particular the existence of a maximum, which is the base set E itself. Since if X is a part of E, obviously $X \subseteq E$.

B.1.3. Derived operations in the transcendental $P(E)$

a Dependence: this is a matter of 'translating' the definition into set-theoretic operations:

$$p \Rightarrow q = \sum\{x \,/\, p \cap x \leq q\}$$

giving:

$$X \Rightarrow Y = \mathbf{U}\{Z \,/\, Z \cap X \subseteq Y\}$$

The dependence of Y with regard to X is the union of all the parts Z of E whose intersection with X is contained in Y. As such, this operation is not especially intuitive. We will see why below.

b Reverse. Translating $\neg p = p \Rightarrow \mu$, we have:

$$\neg X = X \Rightarrow \emptyset = \mathbf{U}\{Z \,/\, Z \cap X = \emptyset\}$$

We see that an element of $\neg X$ is an element of $\mathbf{U}\{Z \,/\, Z \cap X = \emptyset\}$, and as such belongs to a Z such that $Z \cap X = \emptyset$. It follows that an element of $\neg X$ does not belong to X (since $Z \cap X = \emptyset$ means that Z has no element in common with X).

Conversely, if $x \notin X$, the subset $\{x\}$ (the singleton of x) is such that $\{x\} \cap X = \emptyset$. Therefore $\{x\} \subseteq \mathbf{U}\{Z \,/\, Z \cap X = \emptyset\}$. Which means that $\{x\} \subseteq \neg X$. From which we infer that $x \in \neg X$.

Ultimately, $\neg X$ is exclusively composed of all the elements of E which do not belong to X. This is what, in set theory, we call the *complement of X in E*, which we write $\mathbf{C}X$.

B.1.4. Properties of the reverse in the transcendental of the parts of a set

a We can show that $X \cap \neg X = \emptyset$, which is the transcription into our set-theoretic model of the general property (valid for every transcendental) $p \cap \neg p = \mu$. In effect, if $x \in \neg X$, $x \notin X$, and the intersection of X and of $\neg X$ is clearly void.

b We can also show that $\neg\emptyset = E$, which in our particular case transcribes the identity $\neg\mu = M$. Since $\emptyset$ has no elements, its reverse, namely its complement in E, is composed of all the elements of E, and is therefore E itself.

Ex.B.1. Show that $\neg E = \emptyset$, which corresponds to $\neg M = \mu$.

The base set E contains all available elements. Therefore its complement cannot contain any: it is the empty set.

c Yet we also have the very special property $\neg\neg X = X$. In fact, $\neg\neg X$ is the set of elements of E which don't belong to $\neg X$. But this is precisely the set of elements that do belong to X, thus X itself.

We noted above that it is generally not the case in a given transcendental that $\neg\neg p = p$. Certainly, we always have $p \leq \neg\neg p$, but not $\neg\neg p \leq p$. The transcendental $P(E)$ is special in this regard.

d Consider the equation $\neg X = \emptyset$. It signifies that $\neg X$ is void. But any element of E which is not an element of $\neg X$ is an element of X. Therefore X contains all the elements of E, and is therefore identical to E. Consequently, in the transcendental $P(E)$, the equation $\neg X = \emptyset$ has E as its only solution. Formally:

$$\neg X = \emptyset \rightarrow X = E$$

This transcribes, in the terms of T, the general implication:

$$(\neg p = \mu) \rightarrow p = M$$

This implication is however not necessarily valid in any transcendental. We certainly have:

$$(\neg p = \mu) \rightarrow \neg\neg p = \neg\mu \qquad \textit{Thus: } \neg\neg p = M$$

We can only draw from this that $p \leq M$, since $p \leq \neg\neg p$. But we cannot draw the equality $p = M$.

e Another singular property of the transcendental $P(E)$. Recall that the SS ('smallest superior') of two elements p and q, noted $p \cup q$, is the envelope of the subset of T composed of p and q, and noted $\{p, q\}$. This is the smallest of the elements which are larger or equal to both p and q. In our special case, the SS of two parts X and Y of E is their union $X \cup Y$, that is, the set which totalizes the elements of X and those of Y.

It is clear that the union of a part of X and its reverse $\neg X$ gives E in its entirety, or that we have $X \cup \neg X = E$. In effect, $\neg X$ is the set of all the elements of E which do not belong to X. United to those elements belonging to X, they obviously give E in its entirety. It is a major ontological principle, taking the form of the excluded middle, that an element x either belongs or does not belong to X.

The equation $X \cup \neg X = E$ transcribes, returning once again to the notation of T, the identity $p \cup \neg p = M$, which is not the case in every transcendental. In this regard our model is once again rather special.

The reverse $\neg X$, in the transcendental $P(E)$, ultimately has three special properties:

	Set-theoretic expression of the property	General expression of the property
1	$X \cup \neg X = E$	$(p \cup \neg p) = M$
2	$\neg\neg X = X$	$\neg\neg p = p$
3	$(\neg X = \varnothing) \rightarrow (X = E)$	$(\neg p = \mu) \rightarrow (p = M)$

These three properties, as the following exercises demonstrate, are equivalent. Two statements are equivalent if each is the consequence of the other. In other words, π and δ are equivalent if π implies δ ($\pi \rightarrow \delta$) and δ implies π ($\delta \rightarrow \pi$). This is why equivalence is noted as a double implication (its most common symbol is $\leftrightarrow$).

A moment of reflection will show to the reader that if we wish to demonstrate the equivalence of three (or more) statements, we can then proceed to demonstrate that the first statement implies the second, the second the third, and so on. Then we show that the last implies the first.

This is exactly what we will do for the following three properties.

Ex.B.2. Show that in a transcendental T, if $p \cup \neg p = M$, then $\neg\neg p = p$.

$\neg p \cup p = M$	Hypothesis
$\neg\neg p \cap (\neg p \cup p) = \neg\neg p \cap M = \neg\neg p$	Application
$(\neg\neg p \cap \neg p) \cup (\neg\neg p \cap p) = \neg\neg p$	Distributivity
$\mu \cup (\neg\neg p \cap p) = \neg\neg p$	$\neg q \cap q = \mu$
$\neg\neg p \cap p = \neg\neg p$	$\mu \cup q = q$
$\neg\neg p \leq p$	Ex.A.2
$p \leq \neg\neg p$	Ex.A.12
$p = \neg\neg p$	Antisymmetry

Ex.B.3. Show that in a transcendental T, if $\neg\neg p = p$, then the equation $\neg p = \mu$ has as its unique solution $p = M$.

$\neg p = \mu$	Proposed equation
$\neg\neg p = \neg\mu$	Consequence
$\neg\neg p = M$	$\neg\mu = M$
$p = M$	Hypothesis $\neg\neg p = p$

Ex.B.4. Show that in a transcendental T, if the equation $\neg p = \mu$ has $p = M$ as its unique solution, then, for any p, we have $p \cup \neg p = M$.

We will show that $\neg(p \cup \neg p)$ is equal to the minimum μ in any transcendental. The result is that if, in a transcendental, every equation of the kind $\neg x = \mu$ has as its unique solution $x = M$, then $p \cup \neg p = M$.

$\neg(p \cup \neg p) = (p \cup \neg p) \Rightarrow \mu$	Def. of $\neg$
$\neg(p \cup \neg p) = \sum\{t \,/\, t \cap (p \cup \neg p) = \mu\}$	Def. of $\Rightarrow$
$\neg(p \cup \neg p) = \sum\{t \,/\, (t \cap p) \cup (t \cap \neg p) = \mu\}$	Distributivity

But the definition of the SS of two elements r and s (the smallest of the elements superior to both r and s) obviously implies that if $r \cup s = \mu$, this is because both r and s are equal to μ. Finally, we get:

$$\neg(p \cup \neg p) = \sum\{t \,/\, (t \cap p) = \mu \text{ and } (t \cap \neg p) = \mu\} \qquad \text{(I)}$$

By the definition of $\Rightarrow$, $p \Rightarrow \mu$ and $\neg p \Rightarrow \mu$ are upper bounds for every element t such that, respectively, $t \cap p = \mu$ and $t \cap \neg p = \mu$. It follows that the elements t which make up the subset whose above formula (I) expresses the envelope are subject to the two inequalities $t \leq (p \Rightarrow \mu)$ and $t \leq (\neg p \Rightarrow \mu)$. Therefore, given the definition of the reverse, to the two inequalities $t \leq \neg p$ and $t \leq \neg\neg p$.

But then, by the definition of the GI $\cap$ of $\neg p$ and $\neg\neg p$, we have $t \leq \neg p \cap \neg\neg p$. However, the GI of an element and its reverse is equal to the minimum (Ex.A.11). Therefore $t \leq \mu$, meaning that $t = \mu$.

Finally, $\neg(p \cup \neg p)$ is, in any transcendental T, equal to the envelope of a subset all of whose terms are equal to μ, thus to the subset $\{\mu\}$. We have already remarked (Ex.A.9) that such an envelope is itself equal to μ.

If our transcendental of reference verifies the property 'the equation $\neg x = \mu$ has as its unique solution $x = M$', since in any transcendental we have $\neg(p \cup \neg p) = \mu$, it follows, in this particular case, that $p \cup \neg p = M$.

A transcendental that verifies any one of the three equivalent properties is called *Boolean*. We have shown that the transcendental constituted by the parts of a set with the order $\subseteq$ is a Boolean transcendental.

B.1.5. Consider a set possessing only a single element, thus, for example, $\{\varnothing\}$ the singleton of $\varnothing$, whose unique element is the void.

What are the 'parts' of this set? There are only two, namely, $\varnothing$ (which is a part of every set, Cf. B.1.2.) and $\{\varnothing\}$, the set itself. In other words, $P\{\varnothing\} = \{\varnothing\{\varnothing\}\}$.

What is the structure of this transcendental? It is effectively reduced to its minimum and its maximum, since $\varnothing$ is the minimum, and $\{\varnothing\}$ the maximum (Cf. B.1.2, again). In essence, we can write this $\{\mu, M\}$, and we then have these (trivial) correlations:

$$\mu \cap M = \mu$$
$$\mu \cap \mu = \mu$$
$$M \cap M = M$$
$$M \Rightarrow \mu = \mu$$
$$M \Rightarrow M = M$$
$$\mu \Rightarrow M = M$$

We also have, as always, $\neg\mu = M$ and $\neg M = \mu$.

This is our 'minimal' transcendental, which is certainly Boolean, since we obviously have

$$M \cup \neg M = M \cup \mu = M$$

and

$$\mu \cup \neg\mu = \mu \cup M = M$$

This transcendental $T = \{\mu, M,\}$, or equally $T = \{\varnothing, E\}$, is in fact *the transcendental of ontology as such* (of set theory, considered as a mathematical situation). This is the 'basic' Boolean algebra.

B.1.6. There are many ways of demonstrating the intimate connection between the ontology of the pure multiple (here, the set of the parts of a set) and Boolean or classical logic. The principal way, which is Stone's famous theorem, is that every Boolean transcendental can be considered as an 'algebra of sets' in a sense we which will clarify momentarily. In essence, algebras on a set (with $\subseteq$, $\cap$, **U**, $\varnothing$, E) and Boolean algebras are equivalent structures.

B.1.7. For the moment, let us introduce a concept of considerable philosophical interest, because it serves to 'generalize' the intuitive notion of *point.*

Take a function f of a transcendental T toward a transcendental T'. We will say that this function is a *homomorphism* if it conserves the conjunction (the GI), $\cap$, and the envelope, $\sum$.

In order that this 'conservation' is clear, we will note the constitutive operations of T' as $\cap'$ and $\sum'$. We should then have:

$$f(p \cap q) = f(p) \cap' f(q)$$
$$f(\textstyle\sum B) = \sum'\{f(p) \,/\, p \in B\}$$

We also often say that f is a $\cap$-$\sum$ function from T to T', in order to indicate that it makes a conjunction correspond to the conjunction of the corresponding values, and the same goes for global union.

Ex.B.5. A homomorphism f (an $\cap$-$\sum$ function) from T to T' conserves the order of the POS's T and T'. In other words: if $p \leq q$ in T, then $f(p) \leq f(q)$ in T'.

We know (Cf. Ex.A.2) that $p \leq q$ equates to $p \cap q = p$. If f is $\cap$-$\sum$, we then have:

$f(p \cap q) = f(p) \cap' f(q)$	Homomorphism
$f(p) = f(p) \cap' f(q)$	$p \leq q$
$f(p) \leq' f(q)$	Definition of GI

Ex.B.6. A $\cap$-$\sum$ function conserves the minimum and the maximum. In other words: $f(\mu) = \mu'$ and $f(M) = M'$.

This is easily deduced from the preceding exercise.

Now take the Boolean algebra of the parts of a set E, thus the 'minimal' or 'basic' Boolean algebra $\{\mu, M\}$, with its classical rules (Cf. B.1.5.). What exactly is a homomorphism (a $\cap$-$\sum$ function) of the algebra $P(E)$ toward the algebra $\{\mu, M\}$?

THEOREM: A $\cap$-$\sum$ function φ of $P(E)$ on $\{\mu, M\}$ is determined by an element x of E in the following sense: there exists one and only one $x \in E$ such that, for any part $A \subseteq E$, we have:

$$x \in A \leftrightarrow \varphi(A) = M$$
$$x \notin A \leftrightarrow \varphi(A) = \mu$$

And if φ and φ' are two different homomorphisms, the elements x and x' which characterize them are different.

a For a function φ, there necessarily exists at least one $x \in E$ such that $\varphi(\{x\}) = M$. Recall that $\{x\} \subseteq E$ is the singleton of x, namely the part of E whose unique element is x.

Let us remark first of all that $\mathbf{U}[\{x\} / x \in E]$, that is, the union of *all* the singletons, is none other than E itself.

If now there exists no x such that $\varphi(\{x\}) = M$, this is because, for every singleton, we have $\varphi(\{x\}) = \mu$. (In fact, since φ takes its values from the minimal algebra (M, μ), it only has two possible values, M and μ.)

But if for all the singletons $\varphi(\{x\}) = \mu$, we have:

$$\sum\{\varphi(\{x\}) / x \in E\} = \sum\{\mu\} = \mu$$

And yet, since φ is a $\cap$-$\sum$ function, we necessarily have:

$$\varphi[\mathbf{U}\{\{x\} / \varphi(\{x\}) = \mu\}] = \sum\{\varphi(\{x\}) / x \in E\}$$

Therefore, to summarize all of this:

$$\varphi(E) = \mu$$

Which contradicts the exercise B.6 above.

We must therefore abandon the initial hypothesis, and there certainly exists at least one x such that $\varphi(\{x\}) = M$.

b There exists only one x such that $\varphi(\{x\}) = M$. The whole point is that, if $x \neq x'$, $\{x\} \cap \{x'\} = \varnothing$ (an obvious but important remark). Yet, because φ is $\cap$-$\sum$, we have: $\varphi[\{x\} \cap \{x'\}] = \varphi(\{x\}) \cap \varphi(\{x'\})$.

Suppose that $\varphi(\{x\}) = \varphi(\{x'\}) = M$. Then $\varphi[\{x\} \cap \{x'\}] = M \cap M = M$. But $\{x\} \cap \{x'\} = \varnothing$. We should therefore have $\varphi(\varnothing) = M$, which again contradicts exercise B.6.

For any function φ which is $\cap$-$\sum$, there exists finally one and only one $x \in E$ such that $\varphi(\{x\}) = M$.

c Now take a given part A of E, and the unique x such that $\varphi(\{x\}) = M$. If $x \in A$, then $\{x\} \subseteq A$. It follows that $\{x\} \cap A = \{x\}$. Consequently:

$$\varphi[\{x\} \cap A] = \varphi(\{x\}) = M$$
$$\varphi(\{x\}) \cap \varphi(A) = M$$
$$M \cap \varphi(A) = M$$
$$\varphi(A) = M$$

If now $x \notin A$, then $\{x\} \cap A = \varnothing$. Consequently:

$$\varphi[\{x\} \cap A] = \varphi(\varnothing) = \mu$$
$$\varphi(\{x\}) \cap \varphi(A) = \mu$$
$$M \cap \varphi(A) = \mu$$
$$\varphi(A) = \mu$$

Ex.B.7. Show that if $\varphi(A) = M$, then we have, for the x such that $\varphi(\{x\}) = M$, $x \in A$.

If $\varphi(A) = \varphi(\{x\}) = M$, then $\varphi(A \cap \{x\}) = M$. It follows, by exercise B.6, that $A \cap \{x\}$ cannot be equal to $\varnothing$. Therefore, A and $\{x\}$ have at least one element in common, and this element can only be x. Thus $x \in A$.

Ex.B.8. Show that if $\varphi(A) = \mu$, then, again for the same x, we have $x \notin A$.

First of all let's note that if $x \in A$, then $\{x\} \subseteq A$
Consequently $\{x\} \cap A = \{x\}$.
If $\varphi(A) = \mu$, we have $\varphi(A \cap \{x\}) = \mu \cap M = \mu$.
Therefore, $A \cap \{x\}$ cannot be equal to $\{x\}$, since $\varphi(\{x\}) = M$.
It follows that $x \notin A$.

Ultimately, the function φ is entirely characterized by the element x such that $\varphi(\{x\}) = M$ (which both exists and is unique), since:

$$(\varphi(A) = M) \leftrightarrow x \in A$$
$$(\varphi(A) = \mu) \leftrightarrow x \notin A$$

We can therefore say that a $\cap$-$\sum$ function of the Boolean transcendental $P(E)$ on the minimal Boolean transcendental $\{\mu, M\}$ is strictly 'identical' to a determinate element of E.

This is a remarkable correspondence between the ontological relation of belonging ($x \in E$), constitutive of the theory of the pure multiple, and 'Boolean' homomorphisms, which are in truth connections between logical systems. Or: between the concept of element and that of function (or regulated relation).

Essentially, there is a correlation between the being of a situation and its connections within appearing, connections that ultimately always depend on the identification of a real point.

This is why we call a $\cap$-$\sum$ function a *point*. And it is perfectly possible to say that an element of the base set E is identified as a point of the transcendental (here Boolean) constructed on the parts of E. The notion of 'point' is therefore like a mediation between belonging ($x \in E$) and inclusion ($A \subseteq E$). We will note the crucial role played by these very singular parts that are the singletons $\{x\}$.

B.1.8. Turning now to the quasi-complete reciprocity between Boolean algebra and the parts of a set, we will introduce a somewhat more general concept than that of the (complete) set of parts of a set E.

We will call *algebra of subsets on E* (or algebra of parts of E) every family of parts of E closed for the operations $\cap$ (finite intersections), **U** (finite or infinite unions) and **C** (complementary, meaning reverse or negation).

'Closed' means that if A and B are in the family, $A \cap B$ is there also; that if A_i is a given set of members of the family, $\mathbf{U}A_i$ is also in the family, and that if A is in the family $\mathbf{C}A$ is there as well.

Such an algebra of parts is certainly a Boolean algebra. We immediately see that $\varnothing$ is the minimum, and E the maximum. Because if A and $\mathbf{C}A$ are in the family, the same goes for $A \cup \mathbf{C}A = E$, and $A \cap \mathbf{C}A = \varnothing$.

However, it can be quite distinct from $P(E)$. Let's take an elementary example: let A be a part of E. Now consider the family $\{A, \mathbf{C}A, \varnothing, E)$. It has four elements.

Ex.B.9. Show that this family is closed for the characteristic operations of the transcendental.

The unions only give us the elements of the family, since $A \cup \mathbf{C}A = E$, $A \cup E = E$, $A \cup \varnothing = A$, etc.

The intersections do the same, because $A \cap E = A$, $A \cap \varnothing = \varnothing$, $\mathbf{C}A \cap A = \varnothing$, etc.

The above example is that of a finite Boolean transcendental, although it is distinct from the minimal algebra, which would be the family $\{\varnothing, E\}$. More generally we can 'create' an algebra of parts by selecting a family A_i of parts and adding to it all the complementaries of A_i, then all the finite intersections, then all the unions, and so on. It is possible that the family A_i gives us once again the algebra $P(E)$.

Ex.B.10. Show that if we initially select all the singletons of E, we obtain the transcendental $P(E)$.

Any given part A of E is equal to the union of the singletons of the elements which belong to this part. Therefore, any given part A is in the Boolean transcendental that we obtain by starting from the family of all singletons, since we must take all the possible unions in this family. This is to say that all the elements of $P(E)$ are also obtained.

But one can of course obtain, as in exercise B.9, algebras of subsets entirely distinct from $P(E)$, and which moreover may be finite or infinite. Nevertheless we demonstrate:

STONE'S THEOREM: Any Boolean transcendental is isomorphic to an algebra of subsets on a set E.

The (very exciting) demonstration is not required here, but its result indicates in all clarity the conceptual equivalence of which we were speaking.

Between set-theoretic ontology and Boolean logic (and therefore the excluded middle, and the neutralization of double negation), there is an essential reciprocity. 'Boole' is the name of the logic of ontology *as a situation*. This is no way means that Boole names the logic of every situation.

B.2. Connections between the transcendental and logic in its ordinary sense (propositional logic and first order predicate logic)

B.2.1. Consider a simple predicative statement of the kind $P(x)$, which reads: 'x has the property P'. The letter x here designates a variable, an indeterminate term. Regarding such a term, we do not know whether or not it possesses the property in question. This is why an expression like $P(x)$ will be called *open*: its value (true, or false, or probable, or unlikely, etc.) in fact depends on the *determinate* term that we substitute for x. The variable x will be called *free*.

If the letter now designates a determinate term – if 'a' is in fact the proper name of an existent – then we must be able to know if this existent does or does not posses the property P. In which case we say that $P(a)$ is a *closed* expression, and we call a (in contrast to the variable x) a *constant*. This is the basic difference between the contextless sentence 'the thing is white' – where, in the absence of knowing what the thing is, we cannot know its truth-value – and the sentence 'the snow is white', which is true.

Suppose now that we have at our disposal a language with variables (x, y, z, ...), constants (a, b, c, ...) and predicates (P, Q, R). We can interpret the statements constructed in this language in a Transcendental T in the following way:

1. If $P(a)$ is true, we attribute to it the value M (the maximum in T).
2. If $P(a)$ is false, we attribute to it to the value μ (the minimum in T).
3. If there are, in the selected T, elements other than μ and M, say p, then $P(a) = p$ signifies that the statement, being neither true nor false, has an 'intermediate' value, for example 'a probability of being true', or ' true in some particular cases, but generally false', etc.

Thus 'the snow is grey', which is, properly speaking, neither true nor false, since it can be true if the snow is dirty or melting, etc.

The transcendental structure, as described in chapter A, will allow us to interpret the logical connections.

1. What is the value of $[P(a)$ and $Q(b)]$, which involves simultaneously affirming $P(a)$ and $Q(b)$? Intuitively, we know that $[P(a)$ and $Q(b)]$, is true provided that '*a*' possess the property P, and that '*b*' also possesses the property Q. If only one of these clearly does not possess the property – if for example $Q(b)$ is false – then $[P(a)$ and $Q(b)]$ is certainly false. We will generally say that $[P(a)$ and $Q(b)]$ is at least as true as whichever of the two has the lowest truth-value. So, if $P(a)$ is true, but $Q(b)$ is merely probable, their conjunction is only probable.

   It is therefore entirely reasonable to interpret the value of $[P(a)$ and $Q(b)]$ as being, in the transcendental, the GI of the presumed values of $P(a)$ and $Q(b)$. In fact, the conjunction of p and q, that is, $p \cap q$, is the largest of all those which are less than or equal to p and to q. If for example $P(a) = M$ and $Q(b) = M$ (both are true), then $[P(a)$ and $Q(b)]$ are valued $M \cap M = M$. If $P(a) = M$ and $Q(b) = \mu$, then $[P(a)$ and $Q(b)] = M \cap \mu = \mu$ (these were our first two examples above), since M interprets the true and μ the false.

   Now, if $P(a) = M$ (true) and $Q(b) = p$ (probable), then $[P(a)$ and $Q(b)] = M \cap p = p$, since $p \leq M$ in any transcendental: this is our third example.
2. Our procedure is the same for the value of $[P(a)$ or $Q(b)]$, where the 'or' is non-exclusive (both can be true at the same time). We interpret this by the SS $\cup$, which gives us the 'highest' value of each. In effect, if this time $P(a)$ is false and $Q(b)$ is probable, the statement $[P(a)$ or $Q(b)]$ is probable. This verifies that $\mu \cup p = p$ (rule of SS), since $\mu \leq p$.

   We therefore propose that if $P(a)$ equals p and $Q(b)$ equals q:

$[P(a)$ *and* $Q(b)]$ *equals* $p \cap q$
$[P(a)$ *or* $Q(b)]$ *equals* $p \cup q$

The question of implication follows the same pattern. Intuitively, that $P(a)$ implies $Q(b)$ signifies only that the truth of $P(a)$ necessarily entails the truth of $Q(b)$. This point is undoubtedly verified by the term $p \Rightarrow q$ (dependence) of a transcendental.

Ex.B.11. Show in T that if $p \Rightarrow q = M$ and $p = M$, then $q = M$.

$p \Rightarrow q = \sum\{t \,/\, p \cap t \leq q\}$	Def. of $\Rightarrow$
$p \Rightarrow q = M$	Hypothesis
$\sum\{t \,/\, p \cap t \leq q\} = M$	Consequence 1
$p = M$	Hypothesis
$\sum\{t \,/\, p \cap t \leq q\} = \sum\{t \,/\, M \cap t \leq q\}$	Consequence
$\sum\{t \,/\, p \cap t \leq q\} = \sum\{t \,/\, t \leq q\}$	$M \cap t = t$

However, $\sum\{t \,/\, t \leq q\}$, the SS of elements less than or equal to q, is obviously q itself. Hence:

$\sum\{t \,/\, p \cap t \leq q\} = q$	Consequence 2

The comparison of consequences 1 and 2 requires that $q = M$. Which means that from the hypothesis $p \Rightarrow q = M$ (the implication is 'true') and $p = M$ (the antecedent is 'true') follows the conclusion $q = M$ (the consequence is 'true'). This is consistent with the intuitive meaning of implication.

The medieval scholars had already observed that *ex falso sequitur quodlibet* (anything follows from the false), meaning that if $P(a)$ is false, the implication of $Q(b)$ by $P(a)$ is always true, whatever $Q(b)$ may be.

This is also valid in a transcendental.

Ex.B.12. Show in T that if $p = \mu$, then $p \Rightarrow q = M$, whatever q may be.

$\mu \Rightarrow q = \sum\{t \,/\, t \cap \mu \leq q\}$	Def. of $\Rightarrow$
$t \cap \mu = \mu$	Property of $\cap$
$\mu \Rightarrow q = \sum\{t \,/\, \mu \leq q\} = M$	$\mu \leq q$ is always true

In order to cover the general case, we will simply suppose that the value '$P(a)$ implies $Q(b)$' is $p \Rightarrow q$, if p is the value $P(a)$, and q that of $Q(b)$. Let's say (transcendental) dependence interprets (logical) implication.

There still remains the question of negation, namely the value 'not-$P(a)$'. It is really rather instructive to think that not-$P(a)$, as negative assertion, means that $P(a)$ only implies the false. The value of not-$P(a)$ will therefore be $p \Rightarrow \mu$, if p is the value of $P(a)$. It is thus the reverse of p, according to its (derived) definition in a transcendental T. Let's say that the signification of not-$P(a)$ requires that its evaluation in the transcendental be the reverse of the evaluation of $P(a)$.

We note that it is signification, and not direct evaluation, that is at work here. Not-$P(a)$ in the sense: $P(a)$ implies the false. But this says nothing of the effective truth-value of not-$P(a)$. The statement '$P(a)$ implies the false' can obviously be itself false. This is certainly the case, for example, if $P(a)$ is true. The evaluation of not-$P(a)$ will therefore depend on that of $P(a)$. And in fact, if p is the value of $P(a)$ in T, the value of not-$P(a)$ – which is $p \Rightarrow \mu$ – obviously depends on p. For example, if $p = \mu$ (thus if $P(a)$ is false), then not-$P(a)$ has the value $\mu \Rightarrow \mu$, namely M. Which means that not-$P(a)$ is true.

Here we encounter once again the previously established laws: $\neg \mu = M$, and $\neg M = \mu$, now endowed with a logical signification.

B.2.2. Let us enhance our language by admitting expressions such as 'there exists an x such that $P(x)$', or 'at least one x has the property P', expressions which are frequently formalized by $(\exists x)\ (Px)$.

How is this to be interpreted in a transcendental T? For all the constants of our language, $a, b, c, \ldots$ and for a predicate P, we have in T truth-values corresponding to $P(a), P(b), P(c), \ldots$ These values form a subset, noted A_p of T (with $A_p \subseteq T$). In other words, for all of our determinate terms designated by the proper names $a, b, c, \ldots$ (the constants), we know whether they have the property P, or don't have it, or might have it, or probably don't have it, etc. And these values constitute a subset of T, which is A_p.

Consider the envelope of A_p, that is, $\sum A_p$. It designates the smallest of the elements of T greater than or equal to all the elements of A_p Thus it designates the 'maximal' value attributable in T to the statements $P(a), P(b)$, etc. We can say that $\sum A_p$ designates a truth-value 'at least as great' as all those assigned to $P(a), P(b)$, etc. And consequently that there exists an x which has the property P to the measure of $\sum A_p$, or 'to the degree fixed' by $\sum A_p$.

We will therefore maintain that 'there exists an x such that $P(x)$', has $\sum A_p$ as its value in T, since for every a substituted for x, we have: the value $P(a) \leq \sum A_p$.

This means: $\sum A_p$ has 'at least as much' truth as that assigned to $P(a)$, if a is the constant 'most' appropriate to P.

If, in particular, there exists an a which possesses the property P absolutely, or if a designates an entity which possesses the property designated by P, this means that $P(a)$ is true, therefore that $P(a) = M$, M belongs to A_p (which is the collection of all the values of statements of the kind $P(a)$). But then $\sum A_p = M$, which is, in T, the interpretation of '$(\exists x)$ $P(x)$ is true'. In other words, in this interpretation, the existential judgment is true.

Thus we see (in what is still a still very informal manner) how the existential quantifier $\exists$ may be interpreted as an envelope within the transcendental. Which can also be said: the value of 'there exists an x' is the envelope whose territory is the domain of interpretation for x.

We finally have at our disposal a possible projection of the logical connectives into a transcendental: *and* (conjunction), *or* (disjunction), *implies* (implication), *not* (negation) and *there exists* (existential quantifier). These projections are themselves correlated with assignations of truth-value to elementary statements (also called *atomics*) of the kind $P(a)$. These values necessarily include μ (the false) and M (the true), but they may also include others, depending on the transcendental in question. Grasped according to this function, T can be called a *logical space*.

The case in which T is reduced to μ and to M (minimal Boolean transcendental, Cf. B.1.5) is only the 'classical' case, where the sole truth-values are the true and the false. This minimal Boolean algebra is the most traditional logical space, wherein the excluded middle is validated together with the equivalence of double negation and affirmation.

Is this to say that the space of every 'classical' logic is the minimal algebra $\{M, \mu\}$? No. What matters is actually that the transcendental is 'Boolean', that it verifies the three equivalent properties of B.1.4, even if it contains values other than the true (M) and the false (μ).

By contrast, a transcendental which is not Boolean constitutes a non-classical logical space.

B.2.3. We still haven't addressed the question of universal statements of the type 'every x has the property P', in other words $(\forall x)$ $P(x)$. This time it is a matter of interpreting the universal quantifier.

Consider a subset A of elements of $T (A \subseteq T)$, and let the set B be defined as follows: 'all the elements of T which are smaller than all the elements of A'. In other words, if for every element q of A, p is such that $p \leq q$, then p is an element of B. In other words: $B = \{p \;/\; q \in A \rightarrow p \leq q\}$

Since we are in a transcendental, there exists $\sum B$, the envelope of B. This envelope is the smallest element of T which is larger than all the elements of B (which are smaller than all the elements q of A).

Ex.B.13. Show that $\sum B$ is itself smaller than all the elements of A.

Let a be an element of A. Giving:

$a \cap \sum B = \sum\{a \cap p \mid p \in B\}$ Distributivity

But by the definition of B (all the p's smaller than all the elements of A), and since a *is* an element of A, we have, for every p of B, $p \leq a$. From which it is inferred (cf. above, Ex.A.2) that $a \cap p = p$ for every $p \in B$. It then follows that:

$a \cap \sum B = \sum\{p \mid p \in B\} = \sum B$

The result being $\sum B \leq a$ (Ex. A.2)
Therefore $\sum B$ is less than or equal to any element a of A.

So we see that $\sum B$ is definitively the largest of all the elements p of T which are smaller than all the elements of A. We can call this the *global GI* of A. And we have just shown this exists in every transcendental. We will write this ΠA.

Now let us consider the set of values assigned in T to statements of the kind $P(x)$. That is to say the values of $P(a)$, $P(b)$, … for all the constants a, b, …

As earlier, let A_p be this set (such that $A_p \subseteq T$). The element ΠA_p is less than or equal to all the values assigned in T to the statements of the kind $P(a)$, and it is the largest to have that property. This means every statement of the kind $P(a)$ is 'at least as true' as the degree of truth fixed by ΠA_p. In other words, whatever the constant a may be, the statement $P(a)$ has a degree of truth that is at least equal to ΠA_p, and ΠA_p is the largest value to have this property. This means that *all the terms of T* have the property P at least to the degree fixed by ΠA_p.

We can therefore argue that the truth-value of $(\forall x)\, P(x)$ is determined by ΠA_p.

If for example 'for every x, $P(x)$' is absolutely true, this means that $P(a) = M$ irrespective of what the constant a may be. In this case $A_p = \{M\}$ (no $P(a)$ has a value other than M). But then it is obvious that $\Pi A_p = M$, which is the projection onto T of the fact that $(\forall x)\, P(x)$ is true.

If there exists a constant a (be it only one) which is such that $P(a)$ is absolutely false, this means that $P(a) = \mu$. But then $\mu \in A_p$. And as ΠA_p is inferior to every element of A_p, we have $\Pi A_p \leq \mu$, which is to say $\Pi A_p = \mu$. Which is the projection of the fact that 'for every x, $P(x)$' is absolutely false, that a single case (here named by the constant a) refutes universal totalization, the existent named by a not having the property P.

There can of course be intermediary cases, where $\Pi A_p = p$. In this case we will say that 'for every x, $P(x)$' is only true to the degree p (probable or improbable etc.).

Finally, the operator Π (the global GI) is a coherent interpretation of universal quantification.

B.2.4. It is for reasons of formal simplicity that we have only mentioned predicative formulas of the kind $P(x)$. We could equally well have envisaged relational formulas of the kind $\mathfrak{R}(x, y)$, such as: 'x is situated to the right of y'. It is possible to give a value to this type of statement in a transcendental on the basis of constants. One will ask, for example, what is the value of $\mathfrak{R}(a, b)$. We can then use the operations $\cap$, $\sum$ (for example), to give a value to 'and' or to 'there exists'.

The absolutely general and non-naïve presentation of these kinds of things, employing relations with n terms, preoccupies many logicians. It is laden with extraordinarily tedious subtleties of notation, above all when dealing in detail with quantifiers. If in effect we try to establish the protocol of evaluation for formulas of the kind 'for all x's, there exists y such that for all z's we have $\mathfrak{R}(x, y, z)$', namely $(\forall x)(\exists y)(\forall z)\ [\mathfrak{R}(x, y, z)]$, we need to be careful! And so much more if we wish to clarify the case of a given formula. But it can be done.

Our only goal for the time being is to mark out how a transcendental (a Heyting Algebra) is 'readable' as a logical space. We can summarize this as follows:

Logical categories	Correspondent in T
true	M
false	μ
'and' ($\wedge$)	$\cap$
'or' ($\vee$)	$\cup$
implies ($\rightarrow$)	$\Rightarrow$
not-p ($\neg p$)	$p \Rightarrow \mu$ (or $\neg p$)
'there exists' ($\exists$)	$\sum$
'for every' ($\forall$)	Π

B.3. Connection between the transcendental and the general theory of localizations: Topology

B.3.1. This section will first of all show that a topological space – as we will define the concept – possesses the structure of a transcendental. It is a matter of considering it:

- in terms of elements, the open sets of a topology
- in terms of an order-relation, inclusion (as with the case of set theory, treated in section B.1.)

We will then show which conditions a transcendental must obey to be identified as a topology. To do so we will develop the generic concept of 'point', previously defined in B.1.7.

That the logic of appearing is essentially a logic of place is the heart of the matter: to appear is to be *there*.

B.3.2. The notion of localization, in a general sense, can be approached in several ways. One of the more obvious is by distinguishing the interior from the exterior. I am 'there' provided I am interior to what fixes (or names) the 'there' in question. I am not there if I am elsewhere and therefore exterior to that which prescribes the 'there'.

If we refer back to the first transcendental connection (the parts of a set), which we studied in section B.1., we see that this notion of interior is, on the whole, obvious. An element x is 'in' a part A if it belongs to it, and is exterior if it does not belong to it. But what is the part A? It is simply the set of elements which belong to it. There is no difference between the being of A and its 'interiority'.

We could also say: if there is both an interior and an exterior, there must be a 'passage' from one to the other, a crossing of a border. In the case of a part A of a set, there is no border, nothing to mediate between 'belonging to A' and 'not belonging to A'.

Thus we realize that the basic concept of a part of a set is insufficient for a precise thought of localization. While it certainly provides the underlying multiple-being (the being of 'there' is a part), it doesn't offer any concept of passage, proximity (being 'close to A' doesn't mean anything), border and, ultimately, interior.

To 'topologize' a set E, we will associate to each part of this set its interior. To the part A, we will associate $\mathbf{Int}(A)$. What requisites must this 'interior' obey?

It is clear that the interior of A (with $A \subseteq E$) is 'in' A, therefore that it is a part of A, or that it is included in A. We will posit: $\mathbf{Int}(A) \subseteq A$.

Moreover, once you are within the interior, you are there. That is to say, the interior of the interior is always the interior. This time we will posit: $\mathbf{Int}(\mathbf{Int}(A)) = \mathbf{Int}(A)$.

We can also see that if we have the interior of one part and the interior of another part, what there is in common to these two interiors constitutes the interior of what is common to these two parts. In other words: $\mathbf{Int}(A \cap B) = \mathbf{Int}(A) \cap \mathbf{Int}(B)$.

Lastly, the base set E (of which our A, our B etc. are the parts) is our ontological referent: all the elements of which we speak are initially proposed as belonging to E. Which is to say that our topology operates in E, from the point of view of E. The set E is considered to be 'without exterior,' since it is on it – and within it – that we are defining the concepts of interior and exterior. But if it has no exterior, it would be absurd to endow it with an interior other than itself, taken as its being. We thus propose: $\mathbf{Int}(E) = E$

The axioms of the interior, or topological axioms, will finally be the following:

T1 $\mathbf{Int}(A) \subseteq A$
T2 $\mathbf{Int}(\mathbf{Int}(A)) = \mathbf{Int}(A)$
T3 $\mathbf{Int}(A \cap B) = \mathbf{Int}(A) \cap \mathbf{Int}(B)$
T4 $\mathbf{Int}(E) = E$

We will say that we have defined a topology on E if we have defined an 'interior' function on the set of the parts of E. We can obviously define multiple topologies on the same set E. From these four axioms we draw some immediate properties:

Ex.B.14. Show that $\mathbf{Int}(\varnothing) = \varnothing$.

This is immediately deduced from T1.

Ex.B.15. If $A \subseteq B$, then $\mathbf{Int}(A) \subseteq \mathbf{Int}(B)$.

$A \subseteq B \leftrightarrow A \cap B = A$	$p \leq q \leftrightarrow p \cap q = p$
$\mathbf{Int}(A \cap B) = \mathbf{Int}(A)$	Consequence

$\mathbf{Int}(A \cap B) = \mathbf{Int}(A) \cap \mathbf{Int}(B)$	Axiom T3
$\mathbf{Int}(A) \cap \mathbf{Int}(B) = \mathbf{Int}(A)$	Consequence
$\mathbf{Int}(A) \subseteq \mathbf{Int}(B)$	$p \cap q = p \leftrightarrow p \leq q$

Ex.B.16. The set of parts of E is a topology, if we posit that $\mathbf{Int}(A) = A$, for every part A.

The four axioms are easily verified. This topology will be called a *discrete topology*.

B.3.3. For a given topology, we say that a part is *open*, or is an open set for the topology, if it is identical to its interior.

This designation makes sense: if nothing separates an entity from its interior (if it is 'skinless'), it is indeed open.

This concept is philosophically (Cf. Bergson, Deleuze, Heidegger…) and mathematically very important. To fix its definition (for a given topology on E, thus for a fixed function **Int**):

'A is open' $\leftrightarrow$ $\mathbf{Int}(A) = A$

Some fundamental properties of open sets:

Ex.B.17. E and $\varnothing$ are open.

The first by way of axiom T4, the second through the exercise B.14.

Ex.B.18. The intersection of two open sets is an open set.

Take two open sets O_1 and O_2.

$\mathbf{Int}(O_1 \cap O_2) = \mathbf{Int}(O_1) \cap \mathbf{Int}(O_2)$.	By T3
$\mathbf{Int}(O_1) = O_1$ and $\mathbf{Int}(O_2) = O_2$	Def. of openness
$\mathbf{Int}(O_1 \cap O_2) = O_1 \cap O_2$	Consequence
$O_1 \cap O_2$ is open	Def. of openness

Ex.B.19. The interior of any given part A is an open set.

$\mathbf{Int}(\mathbf{Int}(A)) = \mathbf{Int}(A)$	By T2
$\mathbf{Int}(A)$ is open	Def. of openness

Ex.B.20. The interior of A is the largest open set contained in A.

Let's suppose that, for an open set O_1, we have $O_1 \subseteq A$. By Ex.B.15 we then have: $\mathbf{Int}(O_1) \subseteq \mathbf{Int}(A)$. But since O_1 is open, this means: $O_1 \subseteq \mathbf{Int}\ (A)$. Therefore every open set included in A is included in the interior of A, and $\mathbf{Int}(A)$ is indeed the largest open set included in A. ('Largest' obviously functioning here with regard to the order-relation $\subseteq$.)

In fact, $\mathbf{Int}(A)$ is the union of all the open sets included in A.

THEOREM: The union of any number (even an infinite number) of open sets is an open set.

This property is fundamental, and does not extend to intersections (only the finite intersections of open sets are assuredly open sets, by exercise B.18. above).

Let's demonstrate the theorem. As with B.1.2., let I be an infinite set (called 'set of indices'). And let $O_i \in I$ be a collection of open sets indexed to I. (There are 'I' open sets, each indexed by an element i of I). We write $\cup\ O_i$ for the union of all the open sets of the collection.

Suppose $A = \cup\ O_i$. It follows that, for every i, we have $O_i \subseteq A$. Consequently (by Ex.B.15.), $\mathbf{Int}(O_i) \subseteq \mathbf{Int}(A)$. But since all the O_i's are open sets, we have $\mathbf{Int}(O_i) = O_i$. Finally, for every index i we have $O_i \subseteq \mathbf{Int}(A)$. As a consequence, $\cup\ O_i \subseteq \mathbf{Int}(A)$, or $A \subseteq \mathbf{Int}(A)$. But as we also have (by the axiom T1) $\mathbf{Int}(A) \subseteq A$, we finally get, by antisymmetry, $A = \mathbf{Int}(A)$. And therefore A is an open set.

COROLLARY: Ordered by inclusion $\subseteq$, the open sets of a topology defined on a set E form a transcendental.

a The minimum μ is the empty set $\varnothing$, which is open (Ex.B.17) and from which we know it follows that, for every open set O, $\varnothing \subseteq O$.

b The GI is the intersection $\cap$. Exercise B.18 above indicates that $O_1 \cap O_2$ is open. Whether or not it is the largest open set smaller than O_1 and O_2 (the GI of O_1 and O_2) is trivial. Because, by the definition of the (set-theoretic) intersection, if $O_3 \subseteq O_1$ and $O_3 \subseteq O_2$, then $O_3 \subseteq O_1 \cap O_2$.

c The envelope $\sum A$ is the (possibly infinite) union $\cup$ of all the open sets composing the collection A. The above theorem showed that it is an open set. It is obvious that it is the smallest of those which are larger than all the open sets of the union (thus the envelope of this union). Because if an open set O_p exists such that for all the open sets O_i we have $O_i \subseteq O_p$, then we necessarily have $\cup O_i \subseteq O_p$.

d All that remains is to verify the law of distributivity. In the topological case this is written:

$$O_p \cap (\cup O_i) = \cup (O_p \cap O_i)$$

This is a set-theoretic law that we have already encountered (Cf. B.1.2.). We remain within the open sets, since $O_p \cap O_i$ is open (as are both O_p and O_i), and because the union of these open sets is an open set.

B.3.4. We are thus in possession of a mathematical model of the transcendental: the open sets of a topology on a set E. What matters here is knowing whether, as in the case of parts of a set (section B.1.), we in fact have a Boolean algebra, and thus a purely classical model which validates the three equivalent properties of B.1.4., and particularly the first two, which correspond, in the logical interpretation, to excluded middle ($p \cup \neg p = M$) and the law of the equivalence of double negation and affirmation $\neg\neg p = p$.

To see this clearly, we must return to the question of the reverse of an element which is the transcendental support for the problem of negation.

We know that, in a transcendental, $\neg p = p \Rightarrow \mu$, and that in virtue of the definition of $\Rightarrow$, $\neg p = \sum \{t \,/\, t \cap p = \mu\}$.

Transposed to the case of the open sets of a topology, this definition is written: $\neg O = \cup \{O_i \,/\, O_i \cap O = \varnothing\}$. The reverse (the 'negation') of O is the union of all the open sets whose intersection with O is void.

We will change the notation here. We will write $\tilde{O}$ for the negation of O in a topological transcendental. We will take care to note that $\tilde{O}$ *is not* the complement but its interior. Indeed, 'the largest' part of E to have nothing in common with O is the set-theoretic complement of O, namely the set of all the elements of E which do not belong to O. This is what we used as the reverse in the section B.1. We note it **C**O (set-theoretic complement of O). Every open set whose intersection with O is void is included in **C**O. And reciprocally, the intersection of an open set included in **C**O with O is void.

Finally $\cup \{O_i \,/\, O_i \cap O = \varnothing\}$ is nothing other than the union of all the open sets contained in **C**O. By Ex.B.20, we see that this is just simply *the interior of the set-theoretic complement*. Thus: $\tilde{O} = \mathbf{Int}(\mathbf{C}O)$.

The question is then whether or not this reverse possesses Boolean properties. Doubtless the most distinctive property (concerning the logic of appearing) of a Boolean transcendental is excluded middle, that is, $p \cup \neg p = M$. Which, in the topological notation, becomes: $O \cup \tilde{O} = E$.

But if $O \cup \tilde{O} = E$, as moreover $O \cup \mathbf{C}O = E$ by definition, and $\tilde{O} \subseteq \mathbf{C}O$, we must conclude that $\tilde{O} = \mathbf{C}O$. Only, as we have seen that $\tilde{O}$ is none other than the interior of **C**O, if these are moreover identical, *it is because* **C**O *is an open set*. We have therefore demonstrated the following property: if a transcendental of open sets of a topology is Boolean, this is because the set-theoretic complement of an open set is, for the topology in question, an open set.

The reciprocal is obvious. Because if the set-theoretic complement **C**O of an open set O is an open set, it is clearly the largest open set whose intersection with O is void. **C**O is therefore the reverse of O, and as $O \cup \mathbf{C}O = E$ by definition, the transcendental of open sets is Boolean. Hence:

THEOREM: The transcendental of open sets of a topology is Boolean if and only if the set-theoretic complement of every open set is an open set.

B.3.5. To strengthen the presentation, we will introduce a more expressive terminology, directly articulated upon the interior and the exterior. Given a topology on a set E, we say that a part A of E is *closed* if it is the (set-theoretic) complement of an open set.

Ex.B.21. E and $\varnothing$ are closed.

In fact, $\mathbf{C}E = \varnothing$, which is open, and $\mathbf{C}\varnothing = E$, which is open.

We will note that E and $\varnothing$ are both open and closed. We will call *clopen,* for a topology on E, a part of E which has this property (being closed *and* open). An immediate consequence of the final theorem of B.3.4. is then: if all the open sets of a topology are clopen (thus are also closed), the transcendental of open sets is 'Boolean'.

Suppose that, in a topology, the only open sets are E and $\varnothing$ (meaning that every other part of E is such that $\mathbf{Int}(A) \neq A$). Such a topology is called the *indiscrete (or trivial) topology*. As E and $\varnothing$ are clopen, the algebra of the indiscrete topology is Boolean. In fact, this is the minimal Boolean algebra $\{\mu, M\}$, with two elements, here $\{\varnothing, E\}$, which we listed in B.1.5. The algebra of a discrete topology is also Boolean where $\mathbf{Int}(A) = A$ for every A, since then $\mathbf{Int}(\mathbf{C}A) = \mathbf{C}A$, meaning that for every part A, A and $\mathbf{C}A$ are open, and therefore every part is clopen. This time, we have instead a 'maximal' algebra on $P(E)$, identical to the one we studied in section B.1.

In order for a topology on E to be a non-Boolean transcendental, it suffices that, for a single open set O, we have $\tilde{O} \neq \mathbf{C}O$, which entails that $\mathbf{C}O$ – which is closed by definition – is not open, since it is not identical to its interior $\tilde{O}$. As a consequence, O is not clopen.

The most general topological case is that the open sets are not all clopen. We can therefore say that a topology is in general not a Boolean algebra. It is a non-Boolean (or non-classical) transcendental: it validates neither the excluded middle nor the rule of double negation.

We can now return to the correspondences between transcendentals and topologies:

Topologies	Transcendental T
$\varnothing$	μ
E	M
$\cap$	$\cap$
$\cup$	Σ
Int(*C*O)	$\neg O$

Our three examples of connections show the ubiquity of the concept of the transcendental. It is *ontological,* provided that it designates (in its Boolean form) the structuration of parts of a set. It is *logical,* being a space of evaluation for every given value, be they semantic (the true, the false, or the other truth-values) or syntactical (connectives and quantifiers). It is finally *topological*, since the 'natural' model of this structure is nothing other than the open sets of a topological space.

This triple determination of the concept of transcendental is what allows it to regulate appearing as *localization* (being-there), as *cohesion* (logical form of a being), and as *situation* (underlying multiple-being of being-there).

There is an immanent onto-topo-logical (or 'ontopological') regulation.

B.3.6. It is now a matter of consolidating the general link between the power of localization and the transcendental structure.

The end of section B.3. is rather mathematically involved and may be skipped over by the timid reader. It is however at the heart of the philosophical problem, which is the topological essence of appearing.

We noted (in B.1.7.) that with regard to the Boolean transcendental $P(E)$, we could identify an element of E and some 'points' of the algebra, in the sense that a function φ of $P(E)$ in $\{\mu, M\}$, if it was $\cap$-$\sum$ (or if it was a homomorphism, which conserves the operations GI and envelope), was characterized by an element x of E. We then propose:

DEFINITION: Given any transcendental structure, we call *point* of T a $\cap$-$\sum$ function of T on $\{\mu, M\}$.

In the Boolean transcendental $P(E)$, there are exactly as many points (in the sense of the definition) as there are elements in E, since each singleton $\{x\}$ identifies a point φ. In this case we say that there are 'sufficient points'.

The question of which transcendental structures have (or do not have) sufficient points is a question of considerable philosophical consequence. If we accept that every situation is transcendentally structured according to the logic of its appearing, we will see that the 'real points' that provide support for its creative action are more or less numerous depending on whether T has or does not have 'sufficient points'.

The following procedure actually aims to *build a topology from the points of a transcendental*. In other words, to demonstrate there is always a latent topological structure in T.

Let T be a transcendental structure. And let $\pi(T)$ be the set of its points, that is to say the set of $\cap$-$\sum$ functions φ of T on $\{\mu, M\}$. Let's say that $\varphi \in \pi(T)$ for every homomorphism φ of T on $\{\mu, M\}$.

We will associate to every element p of the transcendental T a set of points, noted Op: the set of points which 'give' p the value M (that is, all φ's such that $\varphi(p) = M$). Formally:

$$Op = \{\varphi \mid \varphi \in \pi(T) \text{ and } \varphi(p) = M\}$$

Op is a set of $\cap$-$\sum$ functions operating between T and $\{\mu, M\}$. Or rather, Op is a subset of the set $\pi(T)$ of all the points of T. We have $Op \subseteq \pi(T)$. If we replace p with another element q of T, we obtain another subset Oq of $\pi(T)$. Construction associates, to every element of T, a part of $\pi(T)$.

THEOREM. Let $\pi(T)$ be the set of points of a transcendental T. And, for every element p of T, let Op be the set of points φ such that $\varphi(p) = M$. By posing, for every part A of $\pi(T)$, $\mathbf{Int}(A) = \cup \{Op \mid Op \subseteq A\}$, we obtain a topology. The Op's are the open sets of this topology.

This theorem extracts from the structure of a given transcendental – which, we recall, is founded solely on the order-relation – a structure of localization (a topology). To do this, we pass from the notion of element, which is still strictly ontological, to that of point. Yet the notion of point is functional: it connects each element, through a function φ, to that matricial form of the Two (the 'yes or no') which is the transcendental minimal $\{\mu, M\}$. The point is a 'solidification' of the element through its homomorphic projection onto the Two. The philosophical commentary on this method is of the highest importance.

We have not given the entire demonstration of the theorem, which is difficult. A few exercises will show how we proceed and the reader can follow through of their own volition, if they so desire.

Ex.B.22. Show that the axiom T1 of topologies, that is, $\mathbf{Int}(A) \subseteq A$, is verified by the 'interior' function defined in the theorem.

By definition, $\mathbf{Int}(A)$ is the union of the Op's which are parts of A. So, every element of $\mathbf{Int}(A)$ is an element of at least one Op possessing this property, and is therefore an element of A. Consequently, $\mathbf{Int}(A)$ is a part of A.

Ex.B.23. Show that the axiom T3 of topologies, that is, $\mathbf{Int}(A \cap B) = \mathbf{Int}(A) \cap \mathbf{Int}(B)$, is verified by the 'interior' function defined in the theorem.

$\mathbf{Int}(A \cap B) = \cup \{Op / Op \subseteq A \cap B\}$. But the *Op*'s which are parts of both *A* and *B* are the *Op*'s which belong both to the interior of *A* and the interior of *B*, since these interiors are the union of the *Op*'s which comprise their parts. It immediately follows that $\mathbf{Int}(A \cap B) = \mathbf{Int}(A) \cap \mathbf{Int}(B)$.

Ex.B.24. Show that the axiom T4 of topologies, that is, $\mathbf{Int}(E) = E$, is verified by the 'interior' function defined in the theorem.

It is a matter of showing that $\mathbf{Int}(\pi(T)) = \pi(T)$.

Let M' be the maximum of *T*. For *every* $\cap$-$\sum$ function φ, thus for every point of *T* (every element of $\pi(T)$), we have $\varphi(M') = M$ (the *M* of $\{\mu, M\}$). Cf. on this point Ex.B.6. But then, $O_{M'}$, which is the set of functions φ such that $\varphi(M') = M$, is nothing other than the totality of $\pi(T)$, since it contains all the functions φ. This $O_{M'}$ is, by definition, a part of the interior of $\pi(T)$, namely (since it is identical to $\pi(T)$), $\pi(T) \subseteq \mathbf{Int}(\pi(T))$. However, Ex.B.22 above shows that $\mathbf{Int}(\pi(T)) \subseteq \pi(T)$. By antisymmetry, we can conclude that $\mathbf{Int}(\pi(T)) = \pi(T)$.

Ex.B.25. Show that the parts *Op* of $\pi(T)$ are the open sets of the topology defined in the theorem.

It is clear that $\mathbf{Int}(Op) = Op$, with regard to the definition of **Int.** Therefore, all the subsets of $\pi(T)$ of the kind *Op* are open sets.

Reciprocally, let *A* be an open set for the topology, thus a part *A* of $\pi(T)$ such that $\mathbf{Int}(A) = A$. We will show that there exists an element *q* of *T* such that $A = Oq$.

That *A* is open, or that $A = \mathbf{Int}(A)$, means: $A = \cup \{Op / Op \subseteq A\}$. In other words, every element φ of *A* belongs to at least one *Op* which is a part of *A*. Formally expressed: $\varphi \in A \leftrightarrow (\exists p)\ [\varphi \in Op \text{ and } Op \subseteq A]$. And $\varphi(p) = M$.

Since T is a transcendental, there exists here the envelope of all the elements p such that Op is a part of A. Let's call this envelope q. We have $q = \sum\{p \,/\, Op \subseteq A\}$. Let φ be an element of A. What is the value of $\varphi(q)$? By the definition of q, we have $\varphi(q) = \varphi[\sum\{p \,/\, Op \subseteq A]$. Which, since φ is $\cap$-$\sum$, finally gives $\varphi(q) = \sum\{\varphi(p) \,/\, Op \subseteq A\}$. Only, we have seen above that if A is open, φ, as element of A, belongs to at least one of the Op's which is a part of A. For this Op, we certainly have (by the definition of Op) $\varphi(p) = M$. Consequently, M belongs to the subset of which the envelope $\sum\{\varphi(p) \,/\, Op \subseteq A\}$ is an upper bound. This means that this envelope is forced to be equal to M. Finally, $\varphi(q) = M$.

Inversely, suppose that for an element φ of $\pi(T)$, and for the envelope q defined above, we have $\varphi(q) = M$. Meaning that, for at least one of the Op's which are parts of A, we have $\varphi(p) = M$. But then, $\varphi \in Op$, which entails that $\varphi \in A$.

Finally, there is a strict equivalence between $\varphi(q) = M$ and $\varphi \in A$. Meaning that A is none other than Oq.

The Op's are therefore the parts X of $\pi(T)$ such that **Int**$(X) = X$, thus the open sets of $\pi(T)$.

To recapitulate: If T is a transcendental structure, and if $\pi(T)$ is the set of its points (the $\cap$-$\sum$ functions of T on $\{\mu, M\}$), then there exists on $\pi(T)$ a topology whose Op's – set of points φ for which $\varphi(p) = M$ – are the open sets.

We therefore pass here from *elements* of T (the p's, q's, etc.) to a topologization of its *points* (the $\cap$-$\sum$ functions on $\{\mu, M\}$), via the indexation of the open sets (the Op's).

B.3.7. We will now define a systematic correlation between the transcendental T and the topological space defined on the points of T (on $\pi(T)$).

Consider the function λ (which operates between T and the open sets Op of the topology on $\pi(T)$) thus defined: $\lambda(p) = Op$. This function is $\cap$-$\sum$, as we will now show.

Ex.B.26. Show that $\lambda(p \cap q) = \lambda(p) \cap \lambda(q)$

By definition, $\lambda(p \cap q) = Op \cap q$.

But $\varphi \in O_{p \cap q}$ which means that $\varphi(p \cap q) = M$, therefore $\varphi(p) \cap \varphi(q) = M$, which requires that $\varphi(p) = \varphi(q) = M$ (table of $\{\mu, M\}$). It therefore follows, by the definition of Op and Oq, that $\varphi \in Op$ and $\varphi \in Oq$. Thus $\varphi \in Op \cap Oq$.

Reciprocally, if $\varphi \in Op \cap Oq$, this is because $\varphi(p) = M$ and $\varphi(q) = M$. It follows that $\varphi(p) \cap \varphi(q) = M$, thus that $\varphi(p \cap q) = M$ (homomorphic character of φ), therefore that $\varphi \in O_{p \cap q}$.

Finally, for every $\varphi \in \pi(T)$, we have $\varphi \in O_{p \cap q} \leftrightarrow \varphi \in Op \cap Oq$. From which it follows that $O_{p \cap q} = Op \cap Oq$. And therefore that $\lambda(p \cap q) = Op \cap Oq$. Which, by definition of $\lambda(\lambda(p) = Op)$, will finally mean that $\lambda(p \cap q) = \lambda(p) \cap \lambda(q)$.

Which shows that the function λ conserves the GI.

Ex.B.27. Show that if $B \subseteq T$, we have $\lambda(\sum B) = \cup\, \lambda(p \,/\, p \in B)$

By definition $\lambda(\sum B) = O_{\Sigma B}$

If $\varphi \in O_{\Sigma B}$, this means that $\varphi(\sum B) = M$. But φ is a point. Therefore $\varphi(\sum B) = \sum\{\varphi(p) \,/\, p \in B\}$. If for every p which belongs to B we had $\varphi(p) = \mu$, the envelope would be equal to μ. But it is equal to M, so there is at least one b of B such that $\varphi(p) = M$. Which, by definition, implies that $\varphi \in Op$. We therefore know that $\varphi \in O_{\Sigma B} \rightarrow (\exists p)\ [p \in B$ and $\varphi \in Op]$.

Inversely, if $\varphi \in Op$ and $p \in B$, we have on the one hand $\varphi(p) = M$, and $\varphi \in O_{\Sigma B}$ on the other ($\sum B$ is always an upper bound of all the elements of B). But then $p \cap \sum B = p$. Hence $\varphi(p \cap \sum B) = \varphi(p) = M$, and so $\varphi(p) \cap \varphi(\sum B) = M$. Which requires that $\varphi(\sum B) = M$, and therefore that we have $\varphi \in O_{\Sigma B}$.

Ultimately, $\varphi \in O_{\Sigma B} \leftrightarrow (\exists p)\ [p \in B$ and $\varphi \in Op]$. Which can also be written: $\lambda(\sum B) = O_{\Sigma B} = \cup\ \{Op \,/\, p \in B\} = \cup\ \lambda(p \,/\, p \in B)$.

Which shows that the function λ conserves the envelopes.

The conservation of GI and envelopes demonstrates the $\cap$-$\sum$ character of the function λ.

To sum up once again: given a transcendental T, there exists on the set of its points $\pi(T)$ a topology whose open sets are the sets Op, which is to say, for each element $p \in T$, the set of points φ such that $\varphi(p) = M$. These open sets themselves constitute a transcendental structure. There also exists a function λ from T toward the set of these open sets which is $\cap$-$\sum$, and which therefore conserves the transcendental structure (GI and envelope).

The later already closely links a transcendental structure to a certain kind of topological space. What kind, exactly? This is what we are going to take a closer look at.

B.3.8. We have noted (in B.1.7.) that in the Boolean transcendental $P(E)$, an element of E is identified by a point $\cap$-$\sum$ (function of $P(E)$ on $\{\mu, M\}$).

There are, we said, 'sufficient points'. What happens when we move to the generality of transcendentals, and, consequently, when we examine, as we have already undertaken, the topology defined on the points of T?

Through a reduplication characteristic of mathematical methodology, we will now consider the points of the transcendental defined by the open sets Op of the topology on $\pi(T)$. As this transcendental was itself constructed from the points of T, we can say that we will examine *the points of the points*, or, more precisely, the points of the space of points. It is as if, starting at T (the transcendental of a given situation), we are trying to determine the deeper strata, points, topology on the points, then points of this topology, each time using the operator of dual choice (yes or no, μ or M) that is the minimal transcendental – which is also the transcendental of the ontological situation.

Take the functions $\cap$-$\sum$ (or $\cap$-$\cup$, in this instance), which we'll call Ψ, and which range from the open sets Op (defined on $\pi(T)$, the set of points of T) to $\{\mu, M\}$. We can connect these functions to the function λ defined in B.3.6., according to the following schema:

$$\begin{array}{ccccccccc} T & & & & \pi(T) & & & & \{\mu, M\} \\ p & \longrightarrow & (\lambda) & \longrightarrow & Op & \longrightarrow & (\Psi) & \longrightarrow & M \text{ or } \mu \end{array}$$

To an element p of T, λ makes correspond a set Op of points of T, that is, the functions φ from T to $\{\mu, M\}$. Then to an Op, Ψ makes correspond either M or μ. In either case these correspondences are homomorphisms, which preserve GI and envelope. Or adopting the classical notation for the composition of functions: $g \circ f$ means that the function f acts before the function g. If we say that $\theta = \Psi \circ \lambda$, we will note that θ is a function from T to $\{\mu, M\}$, and that this function is also $\cap$-$\sum$, since Ψ and λ, its components, are themselves $\cap$-$\sum$.

We will note that we are constructing a point of T by connecting an operation between T and its points (the function λ) and an operation of the kind 'points of the points'.

How does the point θ 'function'? Obviously if $\theta(p) = M$, then we have the schema:

$$p \longrightarrow (\lambda) \longrightarrow Op \longrightarrow (\Psi) \longrightarrow M$$

So in fact $\Psi(Op) = M$. Reciprocally, if $\Psi \in Op$, $\Psi(Op) = M$, and $\theta(p) = M$.

Moreover, if Ψ_1 *and* Ψ_2 are two different points of $\pi(T)$, *and* $\theta_1 = \Psi_1 \circ \lambda$ and $\theta_2 = \Psi_2 \circ \lambda$, then $\theta_1 \neq \theta_2$. Because $\Psi_1 \neq \Psi_2$ means that there is at least one $p \in T$ such that $\Psi_1(Op) \neq \Psi_2(Op)$. Thus, for this p, $\theta_1(p) \neq \theta_2(p)$.

There is a bi-univocal correspondence between the Ψ's and θ's, in the sense that two different Ψ's necessarily correspond to two different θ's, the correspondence taking place in the very composition of θ, since $\theta = \Psi \circ \lambda$.

But what is θ? It is a *point* of T, thus an *element* of $\pi(T)$, the set of the points of T. And what is Ψ? It is a point of $\pi(T)$. Finally, to every *point* Ψ of $\pi(T)$, we make correspond an *element* θ of $\pi(T)$, as we did in the case of the Boolean transcendental $P(E)$.

DEFINITION: A topological space, thought as transcendental of its open sets, will be called *sober* if every point (every $\cap$-$\sum$ function on $\{\mu, M\}$) is identified by an element of the underlying set.

We have just shown that the topological space of the points of T, namely $\pi(T)$, with Op's as open sets, is sober. In fact, each point Ψ of the space $\pi(T)$ is identified, via θ, by an element p of the underlying set T. It remains to be seen what conditions a given transcendental structure – which we know is linked to $\pi(T)$ by the function λ – formally inherits from the sobriety of $\pi(T)$. This comes down to determining the case where λ is an isomorphism between T and $\pi(T)$, that is, the case where a transcendental T is structurally identical to the sober and topological transcendental that we can easily extract from the set $\pi(T)$ of its points.

DEFINITION: We will say that a transcendental T has sufficient points if $[\lambda(p) = \lambda(q)] \rightarrow p = q$. That is to say, if the set Op of points φ where $\varphi(p) = M$ is equal to the set Oq of points Ψ where $\Psi(q) = M$ *only* if p is equal to q.

Having 'sufficient points' basically means that the points 'associated' to an element (those for whom $\varphi(p) = M$) are, taken as a whole (the subset Op), actually determined by the element p. This is a fundamental connection between ontological elementarity and the topology of appearing. If there are sufficient points, the same open set of the space cannot constitute the appearing of two different elements.

We have just seen that a sober topological space is at base a transcendental that has sufficient points.

Now let's assume that a transcendental has sufficient points. The effects of this property on the function λ between T and the open sets Op of the topology on $\pi(T)$ present themselves as follows:

1. It is surjective (every element Op is affected by λ). This is obvious, since $\lambda(p) = Op$.
2. It is an algebraic homomorphism, since it conserves the operations $\cap$ and $\cup$ (and therefore also the minimum, the maximum, and finally the partial order).
3. If there are sufficient points, this is because $p \neq q \to Op \neq Oq$, or $p \neq q \to \lambda(p) \neq \lambda(q)$. The function λ is thus injective (or bi-univocal, or difference conserving).

It follows from all of this that λ, surjective and injective homomorphism, is an isomorphism between the transcendental T and the transcendental of the open sets of the topology on $\pi(T)$, the set of the points of T. These two transcendentals are structurally identical. Which proves the:

THEOREM: If a transcendental structure has sufficient points, it is isomorphic to the open sets of a sober topological space.

Such a transcendental will be called *spatial*. We can basically say that a spatial transcendental is a (sober) topology. This (provisionally) completes the study of the essential connection between transcendental and topological structures.

NOTE: We can easily anticipate that the organization of appearing, its basic logic in a situation, will be significantly affected by the question of whether the transcendental of this situation is spatial or not. Yet we should be aware that *fundamentally non-spatial* transcendental structures could perfectly well exist, in the sense that, lacking points, they are in no way identifiable as sober topological spaces. This is how certain Boolean transcendentals, specifically 'atomless' Boolean algebras, are *pointless*! We should also note in passing the link made here between the question of spatiality and that of the One (the existence or not of atoms). We will return to this connection later as the basis of what we will call the axiom of materialism.

C. THEORY OF APPEARING AND OBJECTIVITY

C.1. The aim of this section is to show how, assuming a transcendental T for every situation, the transcendental operates on the pure multiples that ontologically compose the situation. First we will define a *function of appearing* through which every multiple is indexed on the transcendental. Then we define what is an *object*, which is a specific figure of appearance. Finally, we define the primitive component of every object, or the elementary action of every apparent objectivity. This is the theory of *atoms*. The axiom 'every atom is real' then takes on the meaning of a non-decomposable connection between the logic of appearing and the ontology of the multiple.

From there, we 'reascend' from appearing to being, by studying how the atomic composition of an object affects the underlying multiple-being of this object. This process culminates in the demonstration that every (real, ontological) part of a multiple A possesses, under certain conditions, a unique envelope.

C.2. Let A be a set (thus a pure multiplicity, a pure form of being as such). We suppose that this multiple A appears in a situation S whose transcendental is T. We will call *function of appearing* an indexing of A on the transcendental T defined as follows: it is a function $\mathbf{Id}(x, y)$, to be read 'degree of identity of x and y', which to every pair $\{x, y\}$ of elements of A makes correspond an element p of T.

The intuitive idea is that the measure of the identity between the elements x and y of A, such as they appear in the situation S, is fixed by the element p of the transcendental which is assigned to the pair $\{x, y\}$

by the function **Id**. The fact that T has an order-structure is what allows this measure of identity (phenomenality) and the comparison of these measures. We can then say that x appears more or less identical, or similar, to y.

For example, if $\mathbf{Id}(x, y) = M$ (M being, by convention, the maximum of T), we will say that x and y are 'as identical as they can be'. From the interior of the situation of which T is the transcendental, this amounts to saying that x and y are absolutely identical. If on the other hand $\mathbf{Id}(x, y) = \mu$ (the minimum), we can say that, relative to T (and therefore to the situation where x and y appear), these two elements are absolutely non-identical. If, finally, $\mathbf{Id}(x, y) = p$, where p is 'intermediate', that is, $\mu < p < M$, we will say that x and y are 'p-identical', or that the measure of their identity is p.

In order for **Id** to support the idea of identity in a coherent way, we impose on it two axioms, which are the same as those which govern the equivalence-relation (cf. **A.1.**). In essence: the degree of identity of x and y is the same as that of y and x (axiom of symmetry); and the GI of the degree of identity of x and y with the degree of identity of y and z remains inferior to the degree of identity of x and z. This is in fact a kind of transitivity: if x is identical to y in the measure p, and y is identical to z in the measure q, x is at least as identical to z as that which indicates the degree which 'conjoins' p and q, namely $p \cap q$.

Formally, we write:

Ax. **Id**.1. $\mathbf{Id}(x, y) = \mathbf{Id}(y, x)$
Ax. **Id**.2. $\mathbf{Id}(x, y) \cap \mathbf{Id}(y, z) \leq \mathbf{Id}(x, z)$

Note that we have left out reflexivity, which is nevertheless the first axiom of the equivalence-relation. This is precisely because we do not want a rigid conception of identity, since while the latter suits the determination of multiple-being as such, it does not suit its appearance, or its localization, which requires that we have degrees of identity and difference.

The axiom of rigid (or reflexive) identity would be written: $\mathbf{Id}(x, x) = M$. It would indicate that an element x is identical to itself in an absolute sense (according to the maximum of T). This is certainly true if we wish to speak of the identity 'in itself' of the pure multiple x. But what matters to us is the appearance of x in the situation S, and therefore the degree according to which the element x appears in this situation. We will accurately measure this degree of appearance of x in S by the value in the transcendental of the function $\mathbf{Id}(x, x)$.

We will call *degree of existence* of x in the set A, and therefore in the situation S, the value taken by the function $\mathbf{Id}(x, x)$ in the transcendental of this situation. Thus existence, for a given multiple, is the degree according to which it is identical to itself *insofar as it appears in the situation*. We will note that existence is relative to a situation and that its concept is that of a measure, or a degree.

The intuitive idea is that the more vigorously a multiple-being x affirms its identity in the situation, the more phenomenal existence it has within it. Hence the reflexive relation of the function of appearing, that is, $\mathbf{Id}(x, x)$, is a good evaluation of a being's power of appearing.

Thus if $\mathbf{Id}(x, x) = M$ (the maximum), we will hold that x exists absolutely (relative to A, and therefore to S and to the transcendental T). Whereas if $\mathbf{Id}(x, x) = \mu$ (the minimum), we can say that x does not exist at all (in this situation), or that x inexists for S.

To symbolize this interpretation we will write $\mathrm{E}(x)$ in place of $\mathbf{Id}(x, x)$, and we will read this expression as 'existence of x', nevertheless bearing in mind that this is in no way an absolute term, since it depends not only on the transcendental T but also on the function of appearing. However, there can obviously be many functions of appearing in a single situation. We will return to this point.

An immediate property of the function of appearing is the following: the degree of identity between two elements x and y is less than or equal to the degree of existence of x and the degree of existence of y. In other words: one cannot be 'more identical' to another than one is to oneself. Or: the force of a relation cannot outweigh the degree of existence of the related terms. This is an immediate consequence of the axioms.

Ex.C.1. Show that $\mathbf{Id}(x, y) \leq \mathbf{E}(y)$.

$\mathbf{Id}(x, y) \cap \mathbf{Id}(y, x) \leq \mathbf{Id}(x, x)$	Axiom **Id**.2
$\mathbf{Id}(y, x) = \mathbf{Id}(x, y)$	Axiom **Id**.1
$\mathbf{Id}(x, y) \cap \mathbf{Id}(x, y) \leq \mathrm{E}(x)$	Consequence and def. of $\mathrm{E}(x)$
$\mathbf{Id}(x, y) \leq \mathrm{E}(x)$	$p \cap p = p$

For the same reasons, $\mathbf{Id}(x, y) \leq \mathrm{E}(y)$. It follows that:

$\mathbf{Id}(x, y) \leq \mathrm{E}(x) \cap \mathrm{E}(y)$	Def. of $\cap$

C.3. Here is a crucial definition, namely, the definition of what an object is. We know at least since Kant that every theory of appearing turns around the conception it proposes of the object as unity of appearance. This is because the word 'object' names what is counted as one in appearing. Our definition of the object and its components will align this phenomenal count-as-one with the ontological count of pure multiples.

DEFINITION: We call *object* a couple (A, **Id**), where A is a multiple of the situation S and **Id** is a function of appearing of A on the transcendental T of the situation.

An object is thus a being (a multiple) taken *according to its appearing* (its indexing on the transcendental).

Every object is decomposable into parts. This is the theory of object-parts, or sub-objects, which we will shortly introduce in what we might call the analytic of objectivity.

DEFINITION: An *object-component* is a function π (for 'partition function') that goes from a being A which constitutes the entire being of an object (A, **Id**) toward the transcendental T. Thus a function which, to every element x of A, makes correspond an element p of the transcendental.

The underlying intuitive idea is that $\pi(x) = p$ measures the degree according to which the element x belongs to the component of A whose characteristic partition function is π, and which we will henceforth note A_π. For example, if for a given $x \in A$ we have $\pi(x) = M$ (the maximum), we can say that x belongs 'absolutely' to the component A_π. If $\pi(x) = \mu$ (the minimum, we will say that x doesn't belong to the component at all. If $\pi(x) = p$, we will say that x p-belongs to A_π: it is in this component of the object (A, **Id**) 'in the measure p'.

In order that the function π supports the idea of the object-component, or sub-object of an object (A, **Id**), in a coherent way, we impose on it the following two axioms:

π.1: $\quad \pi(x) \cap \mathbf{Id}(x, y) \leq \pi(y)$

π.2: $\quad \pi(x) \leq \mathbf{E}(x)$

The first axiom indicates that the degree of belonging of y to an object-component cannot be inferior to its degree of identity to x combined with the degree of belonging of this x. In other words: if x belongs 'strongly' to A_π, and if y is 'very identical' to x, y must itself belong very strongly to A_π.

Let's take some significant specific cases. If x belongs absolutely to the sub-object A_π, this means that $\pi(x) = M$. Suppose that the degree of identity $\mathbf{Id}(x, y)$ of y to x is measured by p. The axiom $\pi.1$ then tells us that $M \cap p \leq \pi(y)$. Which means, since $M \cap p = p$, that $p \leq \pi(y)$. We can indeed verify that the degree of belonging of y to the component A_π, which is measured by $\pi(y)$, must in any case be at least equal to its identity to x, measured by p.

The second axiom signifies that the measure of an element's belonging to a component cannot be superior to that of its own phenomenal existence in the situation. If for example $\mathbf{E}(x) = \mu$, this means that x inexists in A for the situation S. It would then be absurd to say that x belonged vigorously to a component of the object $(A, \mathbf{Id})$. The axiom $\pi(x) \leq \mathbf{E}(x)$ reasonably imposes, in this case, that we have either $\pi(x) \leq \mu$, or $\pi(x) = \mu$ (since μ is the minimum). Which means that x does not belong absolutely to the component A_π.

C.4. We will now define some minimal components for a given object $(A, \mathbf{Id})$. This is, in the order of appearing, the point of the One, beneath which there is no possible appearance.

We call atomic object-component, or simply *atom*, an object-component which, intuitively, has *at most* one element (two absolutely distinct elements cannot belong absolutely to it). The whole point is to correctly encode the partition function π of these components so as to prescribe the atomic simplicity of their composition.

In addition to the above axioms $\pi.1$ and $\pi.2$, which apply to every object-component, the appropriate axiom for atoms is (writing $\alpha(x)$ for the function which defines an atom):

$$\alpha_2: \qquad \alpha(x) \cap \alpha(y) \leq \mathbf{Id}(x, y)$$

This axiom clearly indicates that the 'conjoined' degree of belonging of x and y to the component A_α is a function of the degree of identity of x and y. In particular, if x and y are absolutely distinct, then their degree of identity is nil, which we will here write as $\mathbf{Id}(x, y) = \mu$. It follows that if x belongs absolutely to the component A_α, or in other words if $\alpha(x) = M$, then y does not belong to it at all, that is, $\alpha(y) = \mu$. In fact, by the above axiom we must have $\alpha(x) \cap \alpha(y) \leq \mu$. Which, if $\alpha(x) = M$, means that $\alpha(y) = \mu$. The axiom prescribes that two absolutely distinct elements cannot both belong 'absolutely' to the same atomic component. In this sense, an atom is indeed an elementary component (or marked by the One).

Ex.C.2. Show that, if π is an atomic component, axiom $\pi.2$ is derived from α_2.

$\alpha(x) \cap \alpha(y) \leq \mathbf{Id}(x, y)$	Axiom α_2
$\alpha(x) \cap \alpha(x) \leq \mathbf{Id}(x, x)$	Case where $x = y$
$\alpha(x) \leq \mathbf{E}(x)$	Def. of $\mathbf{E}(x)$

The above exercise allows us, since it is atoms that we are concerned with at the moment, to reduce the axiomatic of the (atomic) components to the two following statements:

$\alpha.1$	$\alpha(x) \cap \mathbf{Id}(x, y) \leq \alpha(y)$
$\alpha.2$	$\alpha(x) \cap \alpha(y) \leq \mathbf{Id}(x, y)$

We will then note that, for $a \in A$, or for a being-component of A (an element of the multiple A *in the ontological sense*), we can define a function $\boldsymbol{a}(x)$ *which is an atom*. All we need to do is pose that, for every $x \in A$, this function associates the transcendental measure of the identity of x and of a. That is: $\boldsymbol{a}(x) = \mathbf{Id}(a, x)$.

To establish that $\boldsymbol{a}(x)$ is the partition function of an object of an atom for the object $(A, \mathbf{Id})$, it is enough to verify the two aforementioned axioms of atoms:

1. $\boldsymbol{a}(x)$ verifies axiom $\alpha.1$

$\boldsymbol{a}(x) \cap \mathbf{Id}(x, y) = \mathbf{Id}(a, x) \cap \mathbf{Id}(x, y)$	Def. of $\boldsymbol{a}(x)$
$\mathbf{Id}(a, x) \cap \mathbf{Id}(x, y) \leq \mathbf{Id}(a, y)$	Axiom $\mathbf{Id}.2$
$\boldsymbol{a}(x) \cap \mathbf{Id}(x, y) \leq \mathbf{Id}(a, y)$	Consequence
$\boldsymbol{a}(x) \cap \mathbf{Id}(x, y) \leq \boldsymbol{a}(y)$	Def. of $\boldsymbol{a}$

2. $\boldsymbol{a}(x)$ verifies axiom $\alpha.2$

$\boldsymbol{a}(x) \cap \boldsymbol{a}(y) = \mathbf{Id}(a, x) \cap \mathbf{Id}(a, y)$	Def. of $\boldsymbol{a}$
$\mathbf{Id}(a, x) \cap \mathbf{Id}(a, y) = \mathbf{Id}(x, a) \cap \mathbf{Id}(a, y)$	Axiom $\mathbf{Id}.1$
$\mathbf{Id}(x, a) \cap \mathbf{Id}(a, y) \leq \mathbf{Id}(x, y)$	Axiom $\mathbf{Id}.2$
$\boldsymbol{a}(x) \cap \boldsymbol{a}(y) \leq \mathbf{Id}(x, y)$	Def. of $\boldsymbol{a}$

In such a case, the atomic *object-component* $\boldsymbol{a}(x)$ is entirely determined in appearing – on the basis of the function of appearing **Id** – by a *being-component*, the element $a \in A$.

If a given atom, defined by a function $\alpha(x)$, is identical to a *unique* atom of the kind $\boldsymbol{a}(x)$ – in other words, if there exists a unique $a \in A$ such that for every $x \in A$ we have $\alpha(x) = \boldsymbol{a}(x) = \mathbf{Id}(a, x)$ – *we will say that the atom $\alpha(x)$ is real.*

A real atom is an object-component (thus an objective part of a multiple appearing in the situation S) which is, on the one hand, an atomic component (it is basic, or non-decomposable), and on the other, is strictly determined by an underlying element $a \in A$, which is its ontological substructure.

At the point of a real atom, being and appearing conjoin under the sign of the One. We will then say that an object $(A, \mathbf{Id})$ is *saturated* if every atomic component of $(A, \mathbf{Id})$ is real.

THESIS OF OBJECTIVITY (OR AXIOM OF MATERIALISM): Every object is saturated. Or: every atom is real.

This fundamental thesis signifies that at the level of the One-effect (of the atomicity of components), every object, thus everything that appears, is determined by its ontological composition, since, for every atomic component $\alpha(x)$ of appearing, there is a real element which prescribes this atom.

The axiom of materialism is directly opposed to the Bergsonian or Deleuzian presupposition of the primacy of the virtual.

Henceforth, unless otherwise indicated, by 'object' $(A, \mathbf{Id})$ we mean a saturated object. We will also understand by 'atom $\alpha(x)$' a real atom. We will designate real atoms, that is to say the functions $\mathbf{Id}(a, x)$ or $\mathbf{Id}(b, x)$, by $\boldsymbol{a}$, $\boldsymbol{b}$ (in bold characters).

That every atom is real means that it is prescribed by an element a of A. This does not entail that two ontologically different elements a and b prescribe two different atoms. 'Atom' is a concept of objectivity, thus of appearing, and the laws of difference here are not the same as those of ontological difference. The following exercise specifies under what circumstances two different elements prescribe the same atom. In essence, the atomic partition functions $\boldsymbol{a}(x)$ and $\boldsymbol{b}(x)$ are the same if the elements a and b have the same degree of existence, and if their degree of identity is precisely the degree of their existence. This is profound result.

Ex.C.3. Show that two real atoms $\boldsymbol{a}$ and $\boldsymbol{b}$ are the same (that is, for every x, $\boldsymbol{a}(x) = \boldsymbol{b}(x)$) if and only if $\mathbf{Id}(a, b) = \mathbf{E}a = \mathbf{E}b$

— Direct proposition. We suppose that $\boldsymbol{a}(x) = \boldsymbol{b}(x)$ for every x:

$\mathbf{Id}(a, x) = \mathbf{Id}(b, x)$	Def. of real atoms
$\mathbf{Id}(a, a) = \mathbf{Id}(b, a)$	Consequence (if $x = a$)
$\mathbf{Id}(a, b) = \mathbf{Id}(b, b)$	Consequence (if $x = b$)
$\mathbf{Id}(a, b) = \mathbf{Id}(b, a)$	Axiom **Id**.1
$\mathbf{Id}(a, a) = \mathbf{Id}(b, b) = \mathbf{Id}(a, b)$	Consequence
$\mathbf{E}a = \mathbf{E}b = \mathbf{Id}(a, b)$	Def. of **E**

— Reciprocal proposition. We suppose that $\mathbf{E}a = \mathbf{E}b = \mathbf{Id}(a, b)$:

$\mathbf{Id}(a, x) \cap \mathbf{Id}(a, b) \leq \mathbf{Id}(b, x)$		Axiom **Id**.2
$\mathbf{Id}(a, x) \cap \mathbf{E}a \leq \mathbf{Id}(b, x)$		Hypothesis ($\mathbf{Id}(a, x) = \mathbf{E}a$)
$\mathbf{Id}(a, x) \leq \mathbf{E}a$		Ex.C.1
$\mathbf{Id}(a, x) \cap \mathbf{E}a = \mathbf{Id}(a, x)$		Ex.A.2
$\mathbf{Id}(a, x) \leq \mathbf{Id}(b, x)$	(I)	Consequence

The exact same calculus can be made by permuting a and b (starting this time with $\mathbf{Id}(b, x) \cap \mathbf{Id}(b, a) \leq \mathbf{Id}(a, x)$). The result of this permutation is:

$$\mathbf{Id}(b, x) \leq \mathbf{Id}(a, x) \qquad \text{(II)}$$

The comparison of (I) and (II) gives, through anti-symmetry: $\mathbf{Id}(a, x) \leq \mathbf{Id}(b, x)$, namely, by the definition of real atoms, $\boldsymbol{a}(x) = \boldsymbol{b}(x)$, or $\boldsymbol{a} = \boldsymbol{b}$.

C.5. Our task now is to induce, on the basis of the transcendental indexing, or the legislation of appearing, and through the real character of atoms, a structuration of being 'itself'. We will establish the intelligibility of a counter-effect of appearing on being, or of being-there on indeterminate being. A kind of retroaction of objectivity (structure of objects (A, **Id**) on pure multiplicity (position of the set A).

The first stage of this process is to define the *compatibility* of two elements of A. The intuitive idea is to topologize the multiple A on the basis of the function of appearing through which it is objectivated and localized in a situation. To do this, we can draw on the equivalent of a distance between two elements a and b which is the measure of their identity in an objective presentation of the kind (A, **Id**). We know that this measure, namely $\mathbf{Id}(a, b)$, remains, in the transcendental T of the situation, inferior to the existence of both a and b (Ex.C.1). This is written: $\mathbf{Id}(a, b) \leq \mathbf{E}(a) \cap \mathbf{E}(b)$. We will say that a and b are compatible if they are, with regard to their degree of identity, 'in the same zone of existence'. This means that their degree of identity is *equal* to the GI of their degrees of existence.

DEFINITION: We say that two elements a and b of set A which is the underlying multiple of an object $(A, \mathbf{Id})$ are compatible, and we will note this $a \ddagger b$, if the degree of identity of a and b is equal to the common part of their existences, that is, to the GI of that which measures their existences. Formally:

$$a \ddagger b \leftrightarrow \{[\mathbf{E}(a) \cap \mathbf{E}(b)] = \mathbf{Id}(a, b)\}$$

As $\mathbf{Id}(a, b) \leq \mathbf{E}(a) \cap \mathbf{E}(b)$ regardless of a and b (cf. Ex.C.1), the true particularity of two compatible elements is fully expressed in the inequality $\mathbf{E}(a) \cap \mathbf{E}(b) \leq \mathbf{Id}(a, b)$.

We will establish a very interesting connection between the compatibility of two elements a and b of a multiple A and the atomic composition of an object of which A is the underlying multiple. This is already an onto-logical relation, or a correlation between the pure presentation of the multiple and the legislation of appearing. It is constructed from the axiom of materialism, which states that every atom is real, and is therefore identified by a unique element of A.

Ex.C.4. Show that a and b are compatible if and only if the GI of their values for the atomic functions prescribed by a and b, namely $\boldsymbol{a}(x)$ and $\boldsymbol{b}(y)$, remains inferior to the degree of identity of x and y. Formally:

$$a \ddagger b \leftrightarrow [\boldsymbol{a}(x) \cap \boldsymbol{b}(y) \leq \mathbf{Id}(x, y)]$$

To clarify the meaning and importance of this correlation, we will note that if $\mathbf{Id}(x, y) = \mu$, therefore if the degree of identity of x and y is minimal (meaning that, in the situation, it is nil, or that x and y are absolutely different), in order that a and b may in any case be compatible, the above statement prescribes that $\boldsymbol{a}(x) \cap \boldsymbol{b}(y) = \mu$. In particular, if x is in the atomic component $\boldsymbol{a}$ 'absolutely', in other words if $\boldsymbol{a}(x) = M$, then y must be absolutely absent from the atomic component $\boldsymbol{b}$, that is, $\boldsymbol{b}(y) = \mu$. Otherwise we would not have $\boldsymbol{a}(x) \cap \boldsymbol{b}(y) = \mu$. This means that if the atomic components $\boldsymbol{a}$ and $\boldsymbol{b}$ are compatible, they cannot accept entirely non-identical elements into their composition. If, by contrast, $\boldsymbol{a}(x) \cap \boldsymbol{b}(y) = M$, this means that $\boldsymbol{a}(x) = \boldsymbol{b}(y) = M$, and therefore that x and y are absolutely elements, respectively, of the atomic components $\boldsymbol{a}$ and $\boldsymbol{b}$. The compatibility of $\boldsymbol{a}$ and $\boldsymbol{b}$

then demands that $\mathbf{Id}(x, y) = M$, that is, the absolute identity of x and y. This means that if $\boldsymbol{a}$ and $\boldsymbol{b}$ are compatible, they can 'absolutely' accept into their composition (ontologically) different elements x and y, provided that these elements are 'absolutely' identical in appearing, or qua beings-there.

Demonstration of Ex.C.4.

– Direct proposition. We suppose the compatibility of a and b, that is, $a \ddagger b$.

$\mathbf{E}(a) \cap \mathbf{E}(b) = \mathbf{Id}(a, b)$	Def. of $a \ddagger b$
$\mathbf{E}(a) \cap \mathbf{Id}(a, x) = \mathbf{Id}(a, x)$	Ex. C.1
$\mathbf{E}(a) \cap \mathbf{Id}(b, y) = \mathbf{Id}(b, y)$	Idem
$\mathbf{Id}(a, x) \cap \mathbf{Id}(b, y) = \mathbf{Id}(a, b) \cap \mathbf{Id}(a, x) \cap \mathbf{Id}(b, y)$	Consequence
$\mathbf{Id}(a, b) \cap \mathbf{Id}(a, x) \leq \mathbf{Id}(b, x)$	Axiom **Id**.2
$\mathbf{Id}(a, x) \cap \mathbf{Id}(b, y) \leq \mathbf{Id}(b, x) \cap \mathbf{Id}(b, y)$	Consequence
$\mathbf{Id}(a, x) \cap \mathbf{Id}(b, y) \leq \mathbf{Id}(x, y)$	Axiom **Id**.2
$\boldsymbol{a}(x) \cap \boldsymbol{b}(y) \leq \mathbf{Id}(x, y)$	Def. of real atoms

– Reciprocal proposition. We suppose that $\boldsymbol{a}(x) \cap \boldsymbol{b}(y) \leq \mathbf{Id}(x, y)$.

$\mathbf{Id}(a, x) \cap \mathbf{Id}(b, y) \leq \mathbf{Id}(x, y)$	Hypothesis and definition
$\mathbf{Id}(a, a) \cap \mathbf{Id}(b, b) \leq \mathbf{Id}(a, b)$	Case where $x = a$ and $y = b$
$\mathbf{E}(a) \cap \mathbf{E}(b) \leq \mathbf{Id}(a, b)$	Def. of **E**
$\mathbf{Id}(a, b) \leq \mathbf{E}(a) \cap \mathbf{E}(b)$	Ex.C.1
$\mathbf{Id}(a, b) = \mathbf{E}(a) \cap \mathbf{E}(b)$	Antisymmetry
$a \ddagger b$	Def. of $\ddagger$

On the basis of compatibility – the internal relation of multiple-being induced through objectivation $(A, \mathbf{Id})$ – we will now define on A an order-relation which is like a projection of the transcendental organization of appearing onto multiple-being.

DEFINITION: We will say that $a \in A$ is less than or equal to $b \in A$, and we will write this $a < b$, if (with regard to the function of appearing **Id** which objectivates A in $(A, \mathbf{Id})$), on the one hand, a is compatible with b, and, on the other, the degree of existence of a is less than or equal to that of b. Formally:

$$a < b \leftrightarrow (a \ddagger b) \text{ and } (\mathbf{E}a \leq \mathbf{E}b)$$

Order combines compatibility, which is an atomic property, and the evaluation of degrees of existence. Ultimately, $a < b$ means that the real atomic component $\boldsymbol{b}$ is 'approximately' the same as the component $\boldsymbol{a}$ (this is what is expressed by compatibility), but that the degree of objective existence of element b is superior.

It is still necessary to verify that this definition of relation < is indeed an order-relation.

Ex.C.5. Verify that the relation < defined above is indeed an order-relation.

This is a matter of demonstrating that the relation obeys the three axioms of order: reflexivity, transitivity and antisymmetry (cf. A.2.).

– Reflexivity is easy, as a is obviously compatible with itself, and obviously $\mathbf{E}a \leq \mathbf{E}a$.

– Transitivity is demonstrated in two stages, the first of which is trivial since the transitivity of the condition $\mathbf{E}a \leq \mathbf{E}b$ immediately follows from the order-relation $\leq$ of the transcendental T itself. The whole point is then finally that of showing how, under the conditions of our hypothesis ($a < b$ and $b < c$), compatibility is transitive, that is to say, if a is compatible with b, and b with c, then a is compatible with c.

Suppose then that we have $a \ddagger b$ and $b \ddagger c$. It follows that:

$\mathbf{E}a \cap \mathbf{E}b = \mathbf{Id}(a, b)$		Def. of $a \ddagger b$
$\mathbf{E}b \cap \mathbf{E}c = \mathbf{Id}(b, c)$		Idem
$\mathbf{E}a \cap \mathbf{E}b \cap \mathbf{E}c = \mathbf{Id}(a, b) \cap \mathbf{Id}(b, c)$	(I)	Consequence
$\mathbf{E}a \leq \mathbf{E}b$		Hypothesis $a < b$
$\mathbf{E}a \cap \mathbf{E}b = \mathbf{E}a$		Consequence
$\mathbf{E}a \cap \mathbf{E}c = \mathbf{Id}(a, b) \cap \mathbf{Id}(b, c)$		Substitution in (I)
$\mathbf{Id}(a, b) \cap \mathbf{Id}(b, c) \leq \mathbf{Id}(a, c)$		Axiom **Id**.2
$\mathbf{E}a \cap \mathbf{E}c \leq \mathbf{Id}(a, c)$		Consequence
$a \ddagger c$		Def. of $\ddagger$

– Antisymmetry. We suppose that $a < b$ and $b < a$, and show that a and b are ontologically identical (qua elements of A). The demonstration will involve the real character of atoms, that is, the *uniqueness* required of the element $a \in A$ which determines an atom.

$\mathbf{E}a \leq \mathbf{E}b$ and $\mathbf{E}b \leq \mathbf{E}a$		Hypothesis, definition of $<$
$\mathbf{E}a = \mathbf{E}b$	(I)	Antisymmetry of $\leq$ (in T)
$a \ddagger b$		Hypothesis, definition of $<$
$\mathbf{E}a \cap \mathbf{E}b = \mathbf{Id}(a, b)$		Def. of $a \ddagger b$
$\mathbf{E}a = \mathbf{Id}(a, b)$	(II)	(I), and $p \cap p = p$
$\mathbf{E}a = \mathbf{E}b = \mathbf{Id}(a, b)$		(I) and (II)
$\boldsymbol{a}(x) = \boldsymbol{b}(x)$		Ex.C.3

Yet if the atomic partition functions $\boldsymbol{a}$ and $\boldsymbol{b}$ are the same, there exists, by virtue of the materialist axiom according to which every atom is real, a *unique* element of A which is its atomic substructure. This means that the elements a and b are identical, or that $a = b$. This completes the proof of antisymmetry.

We can also express the relation $<$ in other forms, and in this way return to the relation of compatibility $\ddagger$. These are the structurations of the pure multiple on the basis of its 'objective' appearing, none of which are foundational. This is entirely consistent, since in 'returning' from appearing to being, we consider what happens to the multiple *to the extent that* it objectivates itself in a situation.

First of all we will show that the relation $a < b$ can take a very simple form, which doesn't directly imply compatibility but rather its relation to existence and to identity. In effect, a is less than or equal to b as soon as its own degree of existence is less than or equal to its degree of identity to b. We say that $a < b$ if a exists 'less' than its identity to b. Here again, the philosophical commentaries can be numerous and profound.

Ex.C.6. Show that $a < b \leftrightarrow \mathbf{E}a \leq \mathbf{Id}(a, b)$.

– Direct proposition. We suppose that $a < b$

$\mathbf{E}a \leq \mathbf{E}b$	Def. of $a < b$
$\mathbf{E}a \cap \mathbf{E}b = \mathbf{E}a$	Consequence
$\mathbf{E}a \cap \mathbf{E}b \leq \mathbf{Id}(a, b)$	$a \ddagger b$ (from the fact that $a < b$)
$\mathbf{E}a \leq \mathbf{Id}(a, b)$	Consequence

– Reciprocal proposition. We suppose that $\mathbf{E}a \leq \mathbf{Id}(a, b)$

$\mathbf{Id}(a, b) \leq \mathbf{E}b$		Ex.C.1
$\mathbf{E}a \leq \mathbf{E}b$	(I)	Hypothesis

$\mathbf{E}a \cap \mathbf{E}b \leq \mathbf{Id}(a, b)$		Hypothesis, and $p \cap p \leq p$
$a \ddagger b$	(II)	Def. of ‡
$a < b$		(I) and (II)

We will note that the definition of < justified in the above exercise no longer includes compatibility. That said, the following important property legitimates the word 'compatibility' in a classical way, in accordance with order: if two elements are lesser than or equal to a third element, they are compatible.

Ex.C.7. Show that if there exists a c such that $a < c$ and $b < c$, then $a \ddagger b$.

$a < c$ and $b < c$	Hypothesis
$\mathbf{E}a \leq \mathbf{Id}(a, c)$ and $\mathbf{E}b \leq \mathbf{Id}(b, c)$	Consequence, by Ex.C.6
$\mathbf{E}a \cap \mathbf{E}b \leq \mathbf{Id}(a, c) \cap \mathbf{Id}(c, b)$	Consequence, Ax. **Id**.1
$\mathbf{Id}(a, c) \cap \mathbf{Id}(c, b) \leq \mathbf{Id}(a, b)$	Ax. **Id**.2
$\mathbf{E}a \cap \mathbf{E}b \leq \mathbf{Id}(a, b)$	Transitivity of ≤
$a \ddagger b$	Def. of ‡

C.6. In order to further conduct the topologization of multiple being, we will define, under certain circumstances, a kind of envelope for a subset of A. The major property of the transcendental expressed by the existence of the envelope, namely that of the certain possibility of the totalization of phenomenal intensities, can therefore be projected in turn, to a certain extent, onto the 'neutral' multiple of ontology.

DEFINITION: Let B be a part of A. We will say that u is a *pre-envelope* of B, and will write this $u = \mathbf{V}A$, if the following two conditions are satisfied:

- $b < u$ for every $b \in B$
- $\mathbf{E}u = \sum\{\mathbf{E}b \,/\, b \in B\}$

The pre-envelope is here constructed as an upper bound in A for the order retroactively induced by appearing and its transcendental regulation (the order <), and at the same time as an envelope in T for the degrees of existence of the elements of B. It is therefore an *ontologico-transcendental* concept. There is in general (for a given part B of A) no reason that this pre-envelope exists. We will see that certain conditions of its existence combine the notions of compatibility and order that we have previously introduced. In doing so, we will see how we can completely transfer such an ontologico-transcendental concept to the level of being. At the end of the day we will in fact establish that every subset that obeys the conditions investigated has an envelope (and not only a pre-envelope). This will be the:

THEOREM. Every subset B of A whose elements are pairwise compatible possess, for the order-relation $<$, an envelope in the usual sense (the smallest of all of the upper bounds of B). This envelope is also the unique pre-envelope of B.

This theorem is mathematically crucial, because it organizes the thought of the relation between being and the transcendental under one of the most original and essential concepts of contemporary algebraic topology, the concept of *sheaf*. We will however demonstrate this without recourse to this kind of machinery. The demonstration will take place in five lemmas.

LEMMA 1. Let $B \subseteq A$. If the elements of B are pairwise compatible, the function of A in T defined by the equation $\pi(x) = \sum\{\mathbf{Id}(b, x) / b \in B\}$ is an atom.

Let us reflect on what this function represents. Being given an element x of set A, it considers all the degrees of identity between x and all the elements of the subset B. In other words, it 'situates' x in relation to B. This situation is synthesized by the envelope which fixes as accurately as possible the maximum of the degrees under consideration. Ultimately, $\pi(x)$ means something like 'the greatest proximal identity between x and the elements of B', or 'the highest degree of identity by which x approaches that which composes B'.

The extreme importance of this construction is that it provides the means for comparing an element to a subset on the basis of the measure of the greater or lesser identity between this element and the elements of the subset under consideration. However, it sets a condition for this comparison leading to a well-determined measure: that the elements of B are pairwise compatible. Which, in view of the definition of compatibility, amounts to saying that, when taken in pairs, the degree of existence of elements cannot exceed their degree of identity. We can therefore affirm that B is an existentially homogenous subset, in the sense that existential discrepancies cannot exceed differences. Nothing exists here whose existence is not regulated by its networks of identity to the other elements. The theorem thus comes down to saying that every existentially homogenous subset admits an envelope, thus a kind of punctual synthesis.

Demonstration of lemma 1. This is a matter of verifying that $\pi(x)$ conforms to the two axioms α.1 and α.2 of the atomic partition functions (cf. C.4).

– Under the constant condition of considering all the elements $b \in B$, we have:

$\mathbf{Id}(x, y) \cap \sum\{\mathbf{Id}(b, x)\} = \sum\{\mathbf{Id}(x, y) \cap \mathbf{Id}(b, x)\}$	Distributivity
$\mathbf{Id}(y, x) \cap \mathbf{Id}(x, b) \leq \mathbf{Id}(b, y)$	Ax. **Id**.1 and **Id**.2.
$\sum\{\mathbf{Id}(x, y) \cap \mathbf{Id}(b, x)\} \leq \sum\{\mathbf{Id}(b, y)\}$	Consequence
$\mathbf{Id}(x, y) \cap \pi(x) \leq \pi(y)$	Def. of π

This last statement is the axiom α.1 (for π).

Under the same condition, for every generic pair (b, b') of elements of the subset B and every generic pair (x, y) of elements of A, we have:

$\sum\{\mathbf{Id}(b, x)\} \cap \sum\{\mathbf{Id}(b', y)\} = \sum[\mathbf{Id}(b', y) \cap \sum\{\mathbf{Id}(b, x)\}]$	Distributivity
$\pi(x) \cap \pi(y) = \sum[\mathbf{Id}(b', y) \cap \sum\{\mathbf{Id}(b, x)\}]$	Def. of π
$\pi(x) \cap \pi(y) = \sum\sum\{\mathbf{Id}(b', y) \cap \mathbf{Id}(b, x)\}$	Distributivity

But we now suppose that the elements of B are pairwise compatible. In other words, irrespective of what the pair (b, b') is, we have in any case $b \ddagger b'$. Yet exercise 3.4 showed us that the compatibility of b and b' leads to the validity of inequality: $\mathbf{Id}(b', y) \cap \mathbf{Id}(b, x) \leq \mathbf{Id}(x, y)$, regardless of the pair of elements b and b' of B. It follows, in light of the preceding results, and bearing in mind that an envelope is the least upper bound of its territory:

$$\pi(x) \cap \pi(y) \leq \mathbf{Id}(x, y)$$

Which is exactly axiom α.2 (for π).

The function π thus verifies the two axioms, and accordingly identifies an atom.

Since $\pi(x) = \sum\{\mathbf{Id}(b, x) \,/\, b \in B\}$ is an atom, the axiom of materialism implies that this atom is real. Therefore there exists an element $a \in A$ which identifies this atom, thus an element a such that $\boldsymbol{a}(x) = \mathbf{Id}(a, x) = \pi(x)$. We will show that this element a is a pre-envelope of B (in the sense of C.6) for the order-relation $<$. We must therefore establish on the one hand that a is an upper bound of B, meaning that for every $b \in B$, we have $b < a$, and on the other that the degree of existence of a is, in T, the envelope of the degrees of existence of b, that is, $\mathbf{E}a = \sum\{\mathbf{E}b \,/\, b \in B\}$

LEMMA 2. Let $a \in A$ be the element which identifies the real atom $\boldsymbol{a}$ corresponding to the atomicity of the partition function $\pi(x) = \sum\{\mathbf{Id}(b, x) \,/\, b \in B\}$. We will show that a is a pre-envelope for the order-relation $<$ of the subset $B \subseteq A$.

– The element a is an upper bound of every b.

$\mathbf{Id}(a, x) = \sum\{\mathbf{Id}(b, x) \,/\, b \in B\}$	Def. of $\boldsymbol{a}$

This equality means that, for a fixed $x \in A$, we have $\mathbf{Id}(b, x) \leq \mathbf{Id}(a, x)$. Because $\mathbf{Id}(a, x)$, that envelope of every $\mathbf{Id}(b, x)$, is their upper bound. In particular, in the case where $x = b$, it follows that:

$\mathbf{Id}(b, b) \leq \mathbf{Id}(a, b)$	
$\mathbf{E}b \leq \mathbf{Id}(a, b)$	Def. of $\mathbf{E}$
$b < a$	Ex.C.6, definition of $<$

– The degree of existence of a is the envelope of the degrees of existence of b.

$\mathbf{E}b \leq \mathbf{Id}(a, b)$		Above
$\mathbf{Id}(a, b) \leq \mathbf{E}b$		Ex.C.1
$\mathbf{E}b = \mathbf{Id}(a, b)$	(I)	Antisymmetry
$\mathbf{Id}(a, x) = \pi(x) = \sum\{\mathbf{Id}(b, x) \,/\, b \in B\}$		Def. of $\boldsymbol{a}$
$\mathbf{Id}(a, a) = \sum\{\mathbf{Id}(b, a) \,/\, b \in B\}$		Case where $x = a$
$\mathbf{E}a = \sum\{\mathbf{Id}(a, b) \,/\, b \in B\}$	(II)	Def. of $\mathbf{E}$ and Ax. $\mathbf{Id}$.1
$\mathbf{E}a = \sum\{\mathbf{E}b \,/\, b \in B\}$		(I) and (II)

– Finally, a is indeed a pre-envelope of B, thus $a = \mathbf{V}B$.

We therefore know that a is a pre-envelope of B, *provided that the elements of B are pairwise compatible*. Indeed, it is in fact only under this condition that $\pi(x)$ is an atom, and that a exists, insofar as it identifies this atom as real (axiom of materialism). There is at this point nothing to prevent us thinking that, even if the elements of B are not pairwise compatible, they are nevertheless, in the sense given above, a pre-envelope for the relation $<$.

Lemma 3 actually prohibits us from thinking this.

LEMMA 3. If a subset B of A admits a pre-envelope, its elements are pairwise compatible.

Let $\mathbf{V}B$ be the pre-envelope of B. By definition, $\mathbf{V}B$ is an upper bound for all the elements of B. For two given elements b and b', we therefore have $b < \mathbf{V}B$ and $b' < \mathbf{V}B$. The elements b and b', having the same third element as an upper bound, are compatible, by the exercise C.7. Thus we have $b \ddagger b'$ for every pair of elements of B.

It remains for us to pass from the pre-envelope to the envelope proper. To do this, we will show that the pre-envelope a, defined above (by the function $\pi(x)$) – when it exists, thus when the elements of B are pairwise compatible – is an envelope. Meaning that it is the smallest of the elements of A which is an upper bound for all the elements of B. The uniqueness of such an envelope is guaranteed by antisymmetry. It follows that, for a B whose elements are pairwise compatible, the pre-envelope not only exists, but is moreover unique.

LEMMA 4. Let B be a subset of A whose elements are pairwise compatible. And let a be the element identifying the atom $\sum\{\mathbf{Id}(b, x) / b \in B\}$ which is a pre-envelope of B. This element, which *determines* a pre-envelope, *is* in fact an envelope of B, in the sense that it is, in A, and for the relation $<$, the least upper bound of B.

Let c be an upper bound of B. For every $b \in B$, we have $b < c$. It follows that:

$b < c$	Hypothesis
$\mathbf{E}b \leq \mathbf{Id}(b, c)$	Ex.C.6
$\sum\{\mathbf{E}b / b \in B\} \leq \sum\{\mathbf{Id}(b, c) / b \in B\}$	Consequence

$\mathbf{Id}(a, c) = \sum\{\mathbf{Id}(b, c) / b \in B\}$	Def. of a. Case where $x = c$
$\sum\{\mathbf{E}b / b \in B\} \leq \mathbf{Id}(a, c)$	Consequence
$\mathbf{E}a = \sum\{\mathbf{E}b / b \in B\}$	$a = VB$
$\mathbf{E}a \leq \mathbf{Id}(a, c)$	Consequence
$a < c$	Ex.C.6

We see that a is lesser than or equal to every supposed upper bound of B. Therefore it is indeed, being itself such an upper bound, the envelope of B. In other words, B is a territory of a for the order-relation $<$.

LEMMA 5. Given a subset B whose elements are pairwise compatible, this subset has only one pre-envelope, which identifies the element a (by the function $\pi(x)$).

Let's suppose that there exists, in addition to a, a second pre-envelope u. By definition of the pre-envelope, we then have for every $b \in B$:

$$b < u$$
$$\mathbf{E}u = \sum\{\mathbf{E}b / b \in B\}$$

The second equation implies that $\mathbf{E}u = \mathbf{E}a$, since an envelope $\sum$, as it is here, is unique (by antisymmetry). For the same reason (uniqueness of every envelope), if we suppose that u differs here from a, u is not an envelope of B for $<$. Which implies that $a < u$, since a, which is the envelope of B, must be lesser than or equal to every upper bound of B, which is u. Consequently, $\mathbf{E}a \leq \mathbf{Id}(a, u)$ (by exercise C.6).

But as $\mathbf{E}u = \mathbf{E}a$, we also have $\mathbf{E}u \leq \mathbf{Id}(a, u)$, and so ultimately $u < a$ (also by Ex.C.6). It follows then, by antisymmetry, that $u = a$. We can only presume that u differs from a, which means that the one and only pre-envelope of B is a.

The entire theorem is demonstrated.

This remarkable result shows how once it is seized by a transcendental logic through its indexing on T, a being finds itself equipped with an immanent structure which is itself a kind of transcendental (order and envelope). Along the way, we also demonstrated the uniqueness of the ontologico-transcendental term, or mediator, that is a pre-envelope.

D. TRANSCENDENTAL PROJECTIONS: THEORY OF LOCALIZATION

D.1. The goal of this chapter is to immerse ourselves even further in the topological determination of appearing, specifying what we mean by the 'local' investigation of an object. To do this, we will be guided by the substantial identity between transcendental structure and topological spaces. We have in fact already seen (cf. section B.3 of Section B) that the concept of transcendental (which, mathematically speaking, is a complete Heyting algebra) is above all a generalization of (sober) topological spaces. We can therefore think an element $p \in T$ as an open set, thus as that which is used to localize a point through its neighbourhood, or to identify the interior of a part. If we now consider a component of object $\pi(x)$, or A_π, we know that it is, in global terms, a function of a multiple A on the transcendental of the situation (cf. Section C). How do we move from the global to the local? We can try to localize this component (or object-part) in the transcendental by considering $\pi(x) \cap p$, that is, 'what $\pi(x)$ equals in p', or 'what $\pi(x)$ and the point p have in common'. This process of localization is clarified when we consider the element p as being an open set of a topological space. We will then speak of the localization of the object-component 'relative' to the open set p. This allows us in some sense to analyse the object-components in terms of a local decomposition. We will also be confronted with the inverse problem (the problem of 'recomposition'): given a local analysis of the components of an object, can we reconstruct the object as a whole? The basic definition of this analytico-synthetic movement is the following:

DEFINITION. We call localization on p (or restriction to p) of the object-component $\pi(x)$ the function $\pi(x) \cap p$.

The fundamental remark for the conduct of this entire argument is then given in the following very simple exercise, which establishes that *every localization of an atom is an atom.*

Ex.D.1. If $\alpha(x)$ is an atomic component of A, and if we write $\alpha \lceil p$ for the function $x \to \pi(x) \cap p$, $\alpha \lceil p$ is also an atom.

We simply need to establish that if $\alpha(x)$ is an atom, $\alpha \lceil p$ verifies the two axioms of atomic components (cf. C.4).

— Axiom α.1. We must show that $(\alpha \lceil p)(x) \cap \mathbf{Id}(x, y) \leq (\alpha \lceil p)(y)$.

$\alpha(x) \cap \mathbf{Id}(x, y) \leq \alpha(y)$	α is an atom, Ax. α.1
$\alpha(x) \cap p \cap \mathbf{Id}(x, y) \leq \alpha(y) \cap p$	Consequence
$(\alpha \lceil p)(x) \cap \mathbf{Id}(x, y) \leq (\alpha \lceil p)(y)$	Def. of $\alpha \lceil p$

— Axiom α.2. We must show that $(\alpha \lceil p)(x) \cap (\alpha \lceil p)(y) \leq \mathbf{Id}(x, y)$

$\alpha(x) \cap \alpha(y) \leq \mathbf{Id}(x, y)$	α is an atom, Ax. α.2
$[\alpha(x) \cap p] \cap [\alpha(y) \cap p] \leq \mathbf{Id}(x, y)$	$q \cap p \leq q$
$(\alpha \lceil p)(x) \cap (\alpha \lceil p)(y) \leq \mathbf{Id}(x, y)$	Def. of $\alpha \lceil p$

We can therefore see that the global analysis of objects – of which real atoms are the foundation at the same time as they assure the connection between the logic of appearing and the mathematics of being – can open onto a local analysis without losing its guiding thread. Localized by an element of the transcendental (by an open set), an atom remains an atom.

By virtue of the axiom of materialism (C.6), every atom is real. Therefore, if α is an atom, there always exists $a \in A$ such that $\mathbf{Id}(a, x) = \alpha(x)$. We can therefore designate every atom of the object $(A, \mathbf{Id})$ with the name, say a, of the element of A which identifies it. In Section C we used the notation $\boldsymbol{a}$ so as not to introduce any confusion between, on the one hand, the partition *function* $\mathbf{Id}(a, x)$, noted $\boldsymbol{a}$ – which is a real atom and therefore a nomination in appearing – and, on the other hand, the *element* a, which is an ontological component of A, and therefore a nomination in being. But we now seek to project onto being the effects of appearing, or to ontologize logic. It is therefore very suggestive to simply designate with the letter a the atom whose ontological substructure is a. We will not forget, however, that from this point on, unless indicated to the contrary, 'a' will mean, by notational convention, the function $\mathbf{Id}(a, x)$. It is under these conditions that we henceforth allow ourselves to say things like 'a is an atom'.

We have just seen that, a being in this sense an atom, $a \lceil p$ is also an atom. The axiom of materialism then demands that there exists $b \in A$ such that $(a \lceil p)(x) = \mathbf{Id}(b, x)$. We will call b *the restriction of a to p*, and will write this $b = a \lceil p$. This notation is onto-topological, it 'topologizes' the multiple a on the basis of transcendental localizations. We should not lose sight of the fact that this notation technically means: $\mathbf{Id}(b, x) = \mathbf{Id}(a, x) \cap p$.

A major connection with the preceding developments involves seeing that the transcendental compatibility between elements of the multiple A (the relation $a \ddagger b$ of C.7) can be defined in terms of *restriction of an element a to the measure of existence of b, and restriction of b to the measure of existence of a.*

To establish this result, which integrates the local analysis of objects into the larger ontologico-transcendental movement of the preceding part, we have need of a whole series of technical results on localizations. The reader would do well to practice them.

Ex.D.2. Show that $(a \lceil p) \lceil q = a \lceil (p \cap q)$ and that $a \lceil \mathbf{E}a = a$.

First demonstration:

$((a \lceil p) \lceil q)(x) = (\mathbf{Id}(a, x) \cap p) \cap q$	Def. of $\lceil$
$\mathbf{Id}(a, x) \cap (p \cap q) = a \lceil (p \cap q)$	Def. of $\lceil$
$((a \lceil p) \lceil q)(x) = a \lceil (p \cap q)(x)$	Consequence
$(a \lceil p) \lceil q = a \lceil (p \cap q)$	Notational convention

Second demonstration:

$a \lceil \mathbf{E}a(x) = \mathbf{Id}(a, x) \cap \mathbf{E}a$	(I)	Def. of $\lceil$
$\mathbf{Id}(a, x) \leq \mathbf{E}a$		Ex.C.1
$\mathbf{Id}(a, x) \cap \mathbf{E}a = \mathbf{Id}(a, x)$	(II)	Ex.A.2

$a \lceil \mathbf{E}a(x) = \mathbf{Id}(a, x)$	(I) and (II)
$a \lceil \mathbf{E}a = a$	Notational convention

Ex.D.3. Show that $\mathbf{E}(a \lceil p) = \mathbf{E}a \cap p$.

Let $b = a \lceil p$. It follows that:

$\mathbf{Id}(b, x) = \mathbf{Id}(a, x) \cap p$	Def. of b
$\mathbf{Id}(b, a) = \mathbf{Id}(a, a) \cap p = \mathbf{E}a \cap p$	Case where $x = a$
$\mathbf{Id}(b, b) = \mathbf{Id}(a, b) \cap p = (\mathbf{E}a \cap p) \cap p = \mathbf{E}a \cap p$	Case where $x = b$, consequence
$\mathbf{E}b = \mathbf{E}a \cap p$	Definitions
$\mathbf{E}(a \lceil p) = \mathbf{E}a \cap p$	$b = a \lceil p$

Ex.D.4. Show that $a \lceil \mathbf{Id}(a, b) = b \lceil \mathbf{Id}(a, b)$

We will posit that $g = a \lceil \mathbf{Id}(a, b)$, and $h = b \lceil \mathbf{Id}(a, b)$. We will also write $a(x)$ for the real atomic function $\mathbf{Id}(a, x)$ and $b(x)$ for the atomic function $\mathbf{Id}(b, x)$. It follows that:

$g(x) = a(x) \cap \mathbf{Id}(a, b) = a(x) \cap a(b)$	
$a(x) \cap a(b) \leq \mathbf{Id}(b, x)$	Axiom α.2
$g(x) \leq \mathbf{Id}(b, x)$	Consequence
$\mathbf{Id}(b, x) = b(x)$	Notational convention
$g(x) \leq b(x)$	Consequence
$g(x) \leq \mathbf{Id}(a, b)$	From the definition of g
$g(x) \leq b(x) \cap \mathbf{Id}(a, b)$	Def. of $\cap$
$g(x) \leq h(x)$	Def. of h

An exactly symmetrical reasoning, where we replace $g(x)$ with $h(x)$ at the start, results in $h(x) \leq g(x)$. By antisymmetry, we have $g(x) = h(x)$. Thus, as we had hoped, $a \lceil \mathbf{Id}(a, b) = b \lceil \mathbf{Id}(a, b)$.

Ex.D.5. Show that $a \lceil (\mathbf{E}a \cap \mathbf{E}b) = a \lceil \mathbf{E}b$.

$a \lceil (p \cap q) = (a \lceil p) \cap q$ Ex.D.2
$a \lceil (\mathbf{E}a \cap \mathbf{E}b) = (a \lceil \mathbf{E}a) \lceil \mathbf{E}b$ Application
$a \lceil \mathbf{E}a = a$ Ex.D.2
$a \lceil (\mathbf{E}a \cap \mathbf{E}b) = a \lceil \mathbf{E}b$ Application

With these technical results concerning localizations, we will, as announced above, redefine the compatibility of two elements of a multiple A on the basis of the localization of each on the degree of existence of the other. Thus we intertwine, between being and appearing, global and local concepts.

Ex.D.6. Show that $a \ddagger b \leftrightarrow [a \lceil \mathbf{E}b = b \lceil \mathbf{E}a]$.

— Direct proposition. If $a \ddagger b$, then $a \lceil \mathbf{E}b = b \lceil \mathbf{E}a$.

$a \lceil \mathbf{Id}(a, b) = b \lceil \mathbf{Id}(a, b)$. Ex.D.4 above
$\mathbf{Id}(a, b) = \mathbf{E}a \cap \mathbf{E}b$ Def. of $\ddagger$
$a \lceil (\mathbf{E}a \cap \mathbf{E}b) = b \lceil (\mathbf{E}a \cap \mathbf{E}b)$ Consequence
$a \lceil \mathbf{E}b = b \lceil \mathbf{E}a$ Ex.D.5

— Reciprocal proposition. If $a \lceil \mathbf{E}b = b \lceil \mathbf{E}a$, then $a \ddagger b$.

$a(x) \lceil \mathbf{E}b = b(x) \lceil \mathbf{E}a$ Hypothesis
$\mathbf{Id}(a, x) \lceil \mathbf{E}b = \mathbf{Id}(b, x) \lceil \mathbf{E}a$ Def. of $a(x)$ and $b(x)$
$\mathbf{Id}(a, b) \lceil \mathbf{E}b = \mathbf{Id}(b, b) \lceil \mathbf{E}a$ Case where $x = b$
$\mathbf{Id}(a, b) \cap \mathbf{E}b = \mathbf{E}b \cap \mathbf{E}a$ Def. of $\lceil$
$\mathbf{Id}(a, b) \cap \mathbf{E}a \cap \mathbf{E}b = \mathbf{E}a \cap \mathbf{E}b$ Consequence
$\mathbf{E}a \cap \mathbf{E}b \leq \mathbf{Id}(a, b)$ Ex.A.2
$a \ddagger b$ Def. of $\ddagger$

D.3. Thus we see that the relation of compatibility, induced on the being of the object by its transcendental constitution, means that each of the *elements* of the underlying multiple, taken according to its phenomenal power of one (its real atomicity) and restricted to the norm of the existence of another, is equal to this other restricted to its own norm of existence.

From here we can give a 'topological' interpretation of the order $a < b$ induced by the objectivation of A. This is a remarkable interpretation: we have the order-relation $a < b$ if and only if the element a is the restriction of the element b to the existential value of a. Or if a is precisely the localization of b on the degree of existence of a.

Ex.D.7. $a < b \leftrightarrow a = b \lceil \mathbf{E}a$.

— Direct proposition. If $a < b$, then $a = b \lceil \mathbf{E}a$

$a \lceil (\mathbf{E}a \cap \mathbf{E}b) = a \lceil \mathbf{E}b$		Ex.D.5
$\mathbf{E}a \leq \mathbf{E}b$		From the fact that $a < b$
$\mathbf{E}a \cap \mathbf{E}b = \mathbf{E}a$		Ex.A.2
$a \lceil \mathbf{E}a = a \lceil \mathbf{E}b$		Consequence
$a = a \lceil \mathbf{E}a$		Ex.D.2
$a = a \lceil \mathbf{E}b$		Consequence
$a \lceil \mathbf{E}b = b \lceil \mathbf{E}a$		$a < b$ and Ex.D.6
$a = b \lceil \mathbf{E}a$		Consequence

— Reciprocal proposition. If $a = b \lceil \mathbf{E}a$, then $a < b$

$a = b \lceil \mathbf{E}a$		Hypothesis
$\mathbf{E}a = \mathbf{E}(b \lceil \mathbf{E}a)$		Consequence
$\mathbf{E}(b \lceil \mathbf{E}a) = \mathbf{E}b \cap \mathbf{E}a$		Ex.D.3
$\mathbf{E}a = \mathbf{E}b \cap \mathbf{E}a$		Consequence
$\mathbf{E}a \leq \mathbf{E}b$	(I)	Ex.A.2

$a = b \lceil \mathbf{E}a$		Hypothesis (once again)
$a \lceil \mathbf{E}b = (b \lceil \mathbf{E}a) \lceil \mathbf{E}b$		Consequence
$a \lceil \mathbf{E}b = b \lceil (\mathbf{E}a \cap \mathbf{E}b)$		Ex.D.2
$\mathbf{E}a \leq \mathbf{E}b$		Above
$\mathbf{E}a \cap \mathbf{E}b = \mathbf{E}a$		Ex.A.2
$a \lceil \mathbf{E}b = b \lceil \mathbf{E}a$		Consequence
$a \ddagger b$	(II)	Ex.D.6
$a < b$		By (I) and (II). Def. of $<$

D.4. We thus see the very close links forged between a *global* approach of the transcendental kind, where the operators are the function of appearing **Id** and the function of existence **E** (which is a special case of the identity function), and a *local* approach, where what operates is the restriction of a component to its 'projection' onto an element (an open set) of the transcendental (considered as a topological space). All this will culminate in the definition of a constitutive relation, this time going from T *toward* the multiple-being A, and not, as with $\mathbf{Id}(x, y)$ or $\pi(x)$, from the multiple (being) toward the transcendental (the legislation of appearing). This kind of function will make the object appear as a sheaf of transcendental determinations 'recomposed' together.

But before we get there, some more technical exercises on the combinatorial restrictions with the order < are necessary.

Ex.D.8. Show that $a \lceil p < a$.

$a \lceil \mathbf{E}(a \lceil p) = a \lceil \mathbf{E}a \cap p$	Ex.D.3
$a \lceil \mathbf{E}(a \lceil p) = (a \lceil \mathbf{E}a) \lceil p$	Ex.D.2
$a \lceil \mathbf{E}(a \lceil p) = a \lceil p$	Ex.D.2
$a \lceil p < a$	Ex.D.7

Ex.D.9. If $a < b$, $a \lceil p < b \lceil p$

$b \lceil p \lceil \mathbf{E}(a \lceil p) = b \lceil [p \cap \mathbf{E}(a \cap p)]$	Ex.D.2
$b \lceil p \lceil \mathbf{E}(a \lceil p) = b \lceil [p \cap \mathbf{E}a \cap p]$	Ex.D.3
$b \lceil p \lceil \mathbf{E}(a \lceil p) = b \lceil \mathbf{E}(a \cap p)$	$p \cap p = p$
$b \lceil p \lceil \mathbf{E}(a \lceil p) = (b \lceil \mathbf{E}a) \lceil p$	Ex.D.2
$b \lceil p \lceil \mathbf{E}(a \lceil p) = a \lceil \mathrm{p}$	By $a < b$ and Ex.D.7
$a \lceil p < b \lceil p$	Ex.D.7

Ex.D.10. Show that $a < b \lceil p \leftrightarrow a < b$ and $\mathbf{E}a \leq p$.

1 We will first of all show that $a < b \lceil p \rightarrow a < b$

$a \lceil p < b$	Ex.D.8
$a < b \lceil p$	Hypothesis
$a < b$	Transitivity

2 Next we show that $a < b \lceil p \twoheadrightarrow \mathbf{E}a \leq p$

$a < b \lceil p$	Hypothesis
$\mathbf{E}a \leq \mathbf{E}(b \lceil p)$	Def. of <
$\mathbf{E}(b \lceil p) = \mathbf{E}(b) \cap p$	Ex.D.3
$\mathbf{E}(b) \cap p \leq p$	Def. of $\cap$
$\mathbf{E}a \leq p$	Transitivity of $\leq$

3 Last we show (reciprocally) that ($a < b$ and $\mathbf{E}a \leq p$)$\rightarrow a < b \lceil p$

$(b \lceil p) \lceil \mathbf{E}a = b \lceil (p \cap \mathbf{E}a)$	Ex.D.2
$b \lceil (p \cap \mathbf{E}a) = b \lceil \mathbf{E}a$	Hypothesis $\mathbf{E}a \leq p$
$b \lceil \mathbf{E}a = a$	Hypothesis $a < b$ (via Ex.D.7)

Finally, $(b \lceil p) \cap \mathbf{E}a = a$, which means (Ex.D.7) that $a < b \lceil p$.

Ex.D.11. Show that $a \lceil p = a \leftrightarrow \mathbf{E}a \leq p$.

This is a very important result. It indicates that if the existence of a is of a degree inferior to p, the power of the One of a (its atomic objective projection) is identical to that of a restricted to p, or localized on p.

— Direct proposition. If $\mathbf{E}a \leq p$, then $a \lceil p = a$

$a < a$	Reflexivity
$\mathbf{E}a \leq p$	Hypothesis
$a < a \lceil p$	Ex.D.10, case where $a = b$
$a \lceil p < a$	Ex.D.8
$a = a \lceil p$	Antisymmetry

— Reciprocal proposition. If $a \lceil p = a$, then $\mathbf{E}a \leq p$

$a \lceil p = a$	Hypothesis
$\mathbf{E}a = \mathbf{E}(\mathrm{a} \lceil p)$	Consequence
$\mathbf{E}(\mathrm{a} \lceil p) = \mathbf{E}a \cap p$	Ex.D.3
$\mathbf{E}a = \mathbf{E}a \cap p$	Consequence
$\mathbf{E}a \leq p$	Ex.A.2

To finish these preliminaries, we will demonstrate the coherence between the global analysis and the local analysis of multiples 'seized' in appearing through a very important example: the existence of a pre-envelope for a subset B of a set A. A simple exercise allows us to effectively establish that if B has a pre-envelope, the restriction to p of this pre-envelope is the pre-envelope of the subset constituted by the restrictions to p of all the elements of B. We will basically say that the localization on p of a pre-envelope is the pre-envelope of the localizations on p of the territory of the pre-envelope.

Ex.D.12. Let $B \subseteq A$ with an existing $\mathbf{V}B$ (cf. C.10). Show that $(\mathbf{V}B) \lceil p = \mathbf{V}\{b \lceil p \,/\, b \in B\}$.

Let's posit that $a = \mathbf{V}B$. If follows that, for every $b \in B$:

$b < a$		Def. of **V**
$b \lceil p < a \lceil p$	(I)	Ex.D.9
$\mathbf{E}(b \lceil p) = \mathbf{E}b \cap p$		Ex.D.3
$\sum\{\mathbf{E}(b \lceil p) \,/\, b \in B\} = \sum\{\mathbf{E}b \cap p \,/\, b \in B\}$		Consequence
$\sum\{\mathbf{E}(b \lceil p) \,/\, b \in B\} = p \cap \sum\{\mathbf{E}b \,/\, b \in B\}$		Distributivity (in reverse)
$\mathbf{E}a = \sum\{\mathbf{E}b \,/\, b \in B\}$		Def. of **V**
$\sum\{\mathbf{E}(b \lceil p) \,/\, b \in B\} = p \cap \mathbf{E}a$		Consequence
$\sum\{\mathbf{E}(b \lceil p) \,/\, b \in B\} = \mathbf{E}(a \lceil p)$	(II)	Ex.D.3
$a \lceil p = \mathbf{V}\{b \lceil p \,/\, b \in B\}$		Def. of **V**, by (I) and (II)
$\mathbf{V}B \lceil p = \mathbf{V}(b \lceil p)$		$a = \mathbf{V}B$

D.5. We now come to the transcendental construction of the object, to what might be called its constitutive topology.

This time we will start from the transcendental and 'reascend' toward multiple-being. If we have an object (A, **Id**), it makes sense to associate to an element p of the transcendental *all the elements of A which have the degree of existence p*. We thus proceed toward an analysis of the existence of an object which, in the process of transcendental 'arrangement', carves out from it *disjoint parts whose appearing is homogenous* (all the elements of a part have the same degree of existence). As such, to the maximum M would correspond the part of A composed of all the x's such that $\mathbf{E}x = M$, that is, the x's which exist absolutely in the appearing of A. Or, to μ correspond all the x's of A which inexist in its appearing.

We will call this analysis of the strata of existence of an object its *transcendental functor*. The transcendental functor of the object (A, **Id**), written **F**A, associates to every element p of T the part of A composed of x's such that $\mathbf{E}x = p$. That is, $\mathbf{F}A(p) = \{x \,/\, x \in A \text{ and } \mathbf{E}x = p\}$.

We observe that the correlation guaranteed by **F**A takes place between an *element* of the transcendental and *a subset* of A. The transcendental functor is the schema for a way of thinking that takes hold of objects *analytically*, according to the existential stratification of their appearing. This thought localizes the object (the appearing of the pure multiple) on the basis of, and within, the transcendental.

What happens to the values of the transcendental functor **F**A if $q \leq p$? What is the correlation between **F**$A(p)$ and **F**$A(q)$? This point is fundamental in that it touches on the link between the constitutive relation of the transcendental (order) and the existential analysis of the object. It bears on the relationship between the existentially homogenous strata of the object, viewed from the order-structure of the transcendental.

Let $y \in$ **F**A. By definition, this means: **E**$y = p$. Take the restriction to q of the atom prescribed by y, that is, $y \lceil q$. We have **E**$(y \lceil q) =$ **E**$y \cap q$ (by Ex.D.3) $= p \cap q$ (since y is in the existential stratum p).

Now, if $q \leq p$ then we have $q \cap p = q$. In this case, consequently, **E**$(y \lceil q) = q$, and therefore $(y \lceil q) \in$ **F**$A(q)$.

Finally: if $q \leq p$ then every element of **F**$A(p)$ restricted to q belongs to **F**$A(q)$. By posing that $\varphi_q(y) = y \lceil q$, we get the following commutative diagram:

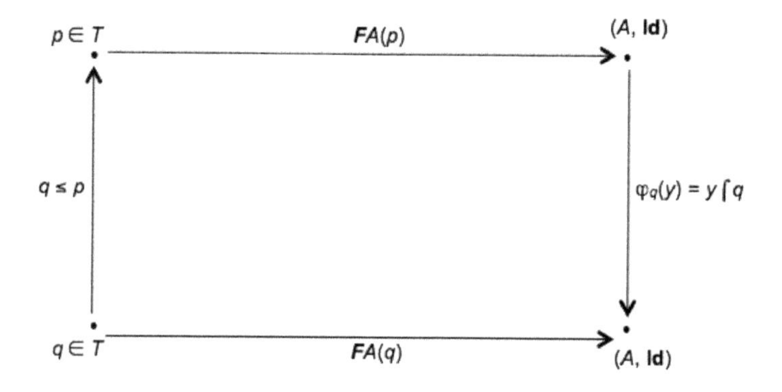

(For advanced readers: this means, in the categorial lexicon, that **F**A *is indeed a contravariant functor between the category of order T and the category of sets.)*

We will see that the functor **F**A has some very remarkable properties. In fact, *it harbours the possibility of a synthesis of existential analysis*. It enacts what we will call a *recomposition* of the transcendental analysis.

D.6. Let us further topologize our examination of the transcendental. In order to do this we will consider the elements of T from a different vantage point than that of their strict belonging to T. We will seize them in their synthetic (or enveloping) function. As we announced in A.4, we will call *territory* of p every part of T of which p is the envelope. The set of the territories of p is written: $\{\Theta \,/\, \Theta \subseteq T \text{ and } p = \sum\Theta\}$.

A territory Θ is that whose global appearing p is capable of measuring.

Our problem is then the following: can we synthesize an existential analysis which is carried out on the basis of a territory in T? Let's suppose that we have a fixed territory Θ for p. Let $(A, \mathbf{Id})$ be an object and $\mathbf{F}A$ its transcendental functor. To each element q of the territory ($q \in \Theta$) corresponds an existential stratum of the object A, such that $\mathbf{F}A(q) = \{x / \mathbf{E}x = q\}$. We could say that the functor associates to the territory Θ a collection of *parts* of the object, a collection of which each member $\mathbf{F}A(q)$ is homogenous in terms of its degree of appearing (or of existence).

We wish to pass analytically from this collection of parts of A to a collection of *elements* of A in such a way that *one* part of the object is associated to the territory Θ. This part would in some sense be the projection of territory Θ into the very being of the object, with the virtue that *each element of the projection would represent a degree of existence.*

Or again, given an element q of the territory, we wish to associate to it *one* element x_q of A, such that $\mathbf{E}x_q = q$ (or $x_q \in \mathbf{F}A(q)$). Since p traverses the territory Θ, the collection of x_q's will be a part of A 'representative' of the territory *according to the existential analysis*, since all the x_q's have different degrees of existence.

Moreover, Θ is a territory *for* p ($p = \sum\Theta$). The analytic projection is only faithful if it is completed by a synthesis which takes place around p, or on the basis of p. But what is the transcendental projection of p? It is the functor $\mathbf{F}A$ applied to p, thus the part of A, noted $\mathbf{F}A(p)$, which brings together all the elements of A whose degree of existence is p. Here again we wish to relate the association of a *part* of A, namely $\mathbf{F}A(p)$, to p (which is the synthesis of the territory Θ), to the association of *a unique* element of A to p. And this element must operate synthetically with regard to all the x_q's, which represent in A the territory Θ.

If we solve this dual problem:

- the selection of x_q's representing the elements of the territory Θ (analysis);
- the determination of a *unique* element $a \in \mathbf{F}A(p)$ which takes a collectivizing position with regard to the x_q's (synthesis);

we will have truly achieved the *transcendental thought of the object*, from a localized point of view (p and its territories).

D.7. In terms of synthesis, the idea clearly results from the examination of the above diagram (D.5). In effect, the correlation between $\mathbf{F}A(p)$ and $\mathbf{F}A(q)$ rests entirely on the function φ, which associates $y \lceil q$ to $y \in$

$\mathbf{F}A(p)$. Let's suppose that our analytic (or representative) problem has been resolved: to each element q of a territory Θ is associated an element x_q of $\mathbf{F}A(q)$. Our synthetic problem will be resolved if we find in $\mathbf{F}A(p)$ a unique element a such that, for every $q \in \Theta$ and every element x_q of $\mathbf{F}A(q)$ associated to q, we have $a \lceil q = x_q$. In other words, we want it to be the case that when we arrange the territory Θ – when $\mathbf{F}A(q)$ is associated to every element q and when an element x_q has been selected in each part $\mathbf{F}A(q)$ – then, for the invariable element a, we get, for every q, $a \lceil q = x_q$.

This means that a carries out in A the recollection of the x_q's, just as, in the transcendental, p carries out the recollection of the territory Θ. Analysis and synthesis are projected, *via* the functors, from the transcendental into the object.

Let's now suppose that we have resolved the problem: such a synthesis exists. For $q \in \Theta$ and $q' \in \Theta$, we have selected x_q and $x_{q'}$ in $\mathbf{F}A(q)$ and $\mathbf{F}A(q')$. By definition, we know that: $\mathbf{E}x_q = q$, $\mathbf{E}x_{q'} = q'$, $a \lceil q = x_q$, $a \lceil q' = x_{q'}$.

But then we have:

$x_q \lceil \mathbf{E}x_{q'} = (a \lceil q) \lceil q' = a \lceil (q \cap q')$	Ex.D.2
$x_{q'} \lceil \mathbf{E}x_{q'} = (a \lceil q') \lceil q = a \lceil (q' \cap q)$	Idem.
$x_q \lceil \mathbf{E}x_{q'} = x_{q'} \lceil \mathbf{E}x_q$	$p \cap q = q \cap p$
$x_q \ddagger x_{q'}$ (x_q and $x_{q'}$ are compatible)	Ex.D.6

This point is crucial: if the synthesis of a territory Θ by a point p of the transcendental is able to be projected into the object, then the elements x_q, which 'represent' the transcendental in the object, are pairwise compatible

Recall (C.7.) that the original definition of compatibility is:

$$\mathbf{E}x_q \cap \mathbf{E}x_{q'} = \mathbf{Id}(x_q, x_{q'})$$

This means that the degree of existence 'common' to x_q and $x_{q'}$ – that is to say their maximal 'common' existence (the GI of x_q and $x_{q'}$) – cannot exceed the measure of their difference. Being compatible means: *two elements coexist as much, and no more, than they differ.* It is only under this condition of compatibility that we can expect to properly solve the problem of transcendental projection.

We will therefore reformulate the problem.

Let Θ be a territory for p. We will call *projective representation of* Θ the association, to every element $q \in \Theta$, of an element x_q of $\mathbf{F}A(q)$ (thus, for an element x_q of A such that $\mathbf{E}x_q = q$). We will say that a projective representation is *coherent* if, for every pair $q \in \Theta$ and $q' \in \Theta$, we have $x_q \ddagger x_{q'}$ (x_q and $x_{q'}$ are compatible).

We will ask whether it is possible, under these conditions, to find in $\mathbf{F}A(p)$ a unique element a (thus $\mathbf{E}a = p$), which takes a synthetic position for a given coherent representation. Meaning that for every $q \in \Theta$, the localization of a in q is x_q, hence $a \lceil q = x_q$.

The answer is positive. The theorem which follows is a kind of recapitulation of all of our efforts, and its philosophical meaning is virtually inexhaustible.

THEOREM: Let A be a set which ontologically underlies an object $(A, \mathbf{Id})$ in a situation S whose transcendental is T. We write $\mathbf{F}A$ for the transcendental functor of A, which associates to every element p of T the subset of A composed of all the elements of A whose degree of existence is p, that is, $\mathbf{F}A(p) = \{x / x \in A$ and $\mathbf{E}x = p\}$. We call territory of p, and write this Θ, every subset of T for which p is the envelope, that is, $p = \sum\Theta$. Finally, we call coherent projective representation of Θ the association, to every element q of Θ, of an element of $\mathbf{F}A(q)$, say x_q (we obviously have $\mathbf{E}x_q = q$) which possesses the following property: for $q \in \Theta$ and $q' \in \Theta$, the corresponding elements of $\mathbf{F}A(q)$ and $\mathbf{F}A(q')$, x_q and $x_{q'}$, are compatible with each other, that is, $x_q \ddagger x_{q'}$. Under these conditions, there always exists one and only one element a of $\mathbf{F}A(p)$ – p being the envelope of Θ – which is such that, for every $q \in \Theta$, the restriction of a to q is uniformly equal to the element x_q of the coherent representation, that is, $a \lceil q = x_q$.

Let's suppose that there exists a coherent projection of a territory Θ for p, and name this projection P. As the elements of P (the x_q's that correspond to the elements q of Θ) are pairwise compatible (definition of coherence), there exists a unique pre-envelope of P, that is, $a = \mathbf{V}P$ (final theorem of Section C)

1 To begin with we show that $\mathbf{E}a = p$, and therefore, as required, $a \in \mathbf{F}A(p)$.
Since a is a pre-envelope of P, we have, by the definition of pre-envelopes: $\mathbf{E}a = \sum\{\mathbf{E}x_q / x_q \in P\}$. But the definition of a projective representation demands that $\mathbf{E}x_q = q$. Therefore, $\mathbf{E}a =$

$\Sigma\{q \,/\, x_q \in P\}$. And since to each $q \in \Theta$ there corresponds one and only one x_q, and if $q \notin q'$ we get $x_q \notin x_{q'}$, it turns out that $\mathbf{E}a = \Sigma\{q \,/\, q \in \Theta\}$. But since p is supposed to have Θ as its territory, we have $\Sigma\{q \,/\, q \in \Theta\} = p$. And therefore, finally, $\mathbf{E}a = p$.

2 We then show that for every $q \in \Theta$, $a \lceil q = x_q$.
Since $a = \mathbf{V}P$, for every $x_q \in P$, we have $x_q < a$. We have in effect established (final theorem of Section C) that the pre-envelope $\mathbf{V}P$ is also the envelope of P for the order $<$. But (Ex.D.7.), $x_q < a$ is also written $x_q = a \lceil \mathbf{E}x_q$, and as $\mathbf{E}x_q = q$ (since $x_q \in \mathbf{F}A(q)$), it turns out that $a \lceil q = x_q$.

Thus there does indeed exist a unique element a such that $\mathbf{E}a = p$, whose restriction to q is, for every q, identical to the selected element x_q, *whatever the selection may be*, provided that the selected elements are pairwise compatible.

Or again, every coherent projective representation of a transcendental territory Θ is synthesized in the pure multiple underlying the appearance of the object by a unique element, whose degree of existence is equal to the transcendental element p of which Θ is the territory.

The being of the object is therefore internally organized, on the basis of the existential analysis, by the syntheses which correspond to the transcendental envelopes (to the territories).

(For advanced readers: we see that, in the categorial vocabulary, the functor **F***A is a sheaf. The category of these sheaves is essentially a Grothendieck topos.)*

E. THEORY OF RELATIONS: SITUATION AS UNIVERSE

Up to this point we have presented the mathematics of the transcendental from the strict point of view of the internal composition of objects, in particular their atomic substructure. We have in effect successfully carried out an analytic of appearing in the framework of the presentational unity constitutive of objectivity. The aim of the present section is to define what *a relation between objects* is, and thereby fully account for the logic of appearing, not only as structural logic of objectivity, but as *Universal logic*, or theory of the relation of objects in a given situation. Sticking with the Kantian lexicon, we could say that it is a matter of sketching a dialectic of objectivity which places the logical givenness of objects within the universe of relations.

The relation will be defined as closely as possible to a kind of ontologico-transcendental invariance or stability. We will posit that a relation is identifiable as relation only to the extent that it 'conserves' the principal transcendental particularities of objects, specifically the degrees of existence and the localizations. Meaning that no relation has the power to disrupt the real atomic substructure of appearing. There is a resistance of matter.

Let us be more precise. Take two objects (A, δ) and (B, γ). (We abandon here the suggestive identitarian notation $(A, \mathbf{Id})$, since we must now designate several distinct objects, therefore several possible functions of appearing. Whence an extensive usage of Greek letters to designate the various functions of appearing assigned to the different objects.) A *relation* from the object (A, δ) to the object (B, γ) is given in appearing if it conserves the fundamental existential and topological properties of this appearing. Otherwise, the relation could not be thought as a relation between *these* objects, and *it would not appear*. It is to the precise extent that it operates through the global maintenance of the transcendental characteristics of the object – and therefore of the logical laws of appearing – that the relation itself appears and can thereby make the situation consist, in appearance, as a universe.

Yet we have already seen that the fundamental existential given associated to an object is the *degree of existence* of the elements of its underlying multiple-being (that is, **E**x), and that the major topological or localizing given is the *restriction* to a transcendental index p of the power of the One (of the atomic projection) of these elements (namely $x \lceil p$). For a connection from multiple A to multiple B to appear as a relation, it is therefore necessary that it retains the degrees of existence and the localizations.

Under this condition, it is simplest to use the old concept of function, namely that which makes an element of set B 'correspond' to every element of set A. If we write $g(x) = y$, we must understand by this that g associates to every x of A one y of B, according to the rules that the singularity of the function must indicate. A relation from (A, δ) to (B, γ) will be a function g from A to B which leaves the existences and the localizations invariant. We will therefore posit that:

DEFINITION: A relation from an object (A, δ) to an object (B, γ) is a function g from the set A to the set B which satisfies, for every $a \in A$,

$$\mathbf{E}g(a) = \mathbf{E}a$$

$$g(a \lceil p) = g(a) \lceil p$$

Let us recall the notational conventions: the letters δ and γ obey the rules for functions of appearing, which were generally written **Id** in the preceding sections. For example, $\gamma[g(a), g(b)]$ designates the degree of identity of elements which, in the multiple B, correspond by the function g to the elements a and b of the multiple A. The transcendental T is presumed to be invariant, since we remain in the same situation S, of which A and B are the multiples.

The conservation of the degrees of existence and the localizations leads to strong consequences for the relation g, in terms of the evaluation of identities.

Ex.E.1. Show that if g is a relation, we have: $\delta(a, b) \leq \gamma[g(a), g(b)]$, for $a \in A$ and $b \in B$.

In other words: a relation cannot *diminish* the degree of identity between two terms. The two correlates $g(a)$ and $g(b)$ are at least as identical in B as a and b are in A. Or again, no relation presents differences greater than the ones initially given. As a rule, identity is conserved or reinforced, never reduced. *A relation does not create difference.*

$$A \lceil \delta(a, b) = b \lceil \delta(a, b) \qquad \text{Ex.D.4}$$
$$g[a \lceil \delta(a, b)] = g[b \lceil \delta(a, b)] \qquad \text{Consequence}$$
$$g(a) \lceil \delta(a, b) = g(b) \lceil \delta(a, b) \qquad \text{Conservation of } \lceil$$

Meaning that, for every y of B, we have:

$$\gamma[g(a), y] \cap \delta(a, b) = \gamma[g(b), y] \cap \delta(a, b) \qquad \text{Def. of } \lceil$$

In particular, for $y = g(a)$, we get:

$$\mathbf{E}g(a) \cap \delta(a, b) = \gamma[g(b), g(a)] \cap \delta(a, b) \qquad \text{Consequence, def. of } \mathbf{E}$$
$$\mathbf{E}a \cap \delta(a, b) = \gamma[g(a), g(b)] \cap \delta(a, b) \qquad \text{Conservation of } \mathbf{E}$$
$$\mathbf{E}a \cap \delta(a, b) = \delta(a, b) \qquad \text{Ex.C.1 and Ex.A.2}$$
$$\delta(a, b) \cap \gamma[g(a), g(b)] = \delta(a, b) \qquad \text{Consequence}$$
$$\delta(a, b) \leq \gamma[g(a), g(b)] \qquad \text{Ex.A.2}$$

In fact, the conservative power of a relation, based on the maintenance of **E** and of $\lceil$, extends to the entire transcendental structuration of the multiple underlying the objects.

Ex.E.2. Show that if a is compatible with b in A – in the sense, written $a \ddagger b$, of Section C – and if g is a relation between (A, δ) and (B, γ) then $g(a)$ is compatible with $g(b)$ in B. That is: $a \ddagger b \rightarrow g(a) \ddagger g(b)$.

We suppose that $a \ddagger b$. It follows that:

$$a \lceil \mathbf{E}b = b \lceil \mathbf{E}a \qquad \text{Hypothesis and Ex.D.6}$$
$$g(a \lceil \mathbf{E}b) = g(b \lceil \mathbf{E}a) \qquad \text{Consequence}$$
$$g(a) \lceil \mathbf{E}b = g(b) \lceil \mathbf{E}a \qquad \text{Conservation of } \lceil \text{ by } g$$
$$g(a) \lceil \mathbf{E}g(b) = g(b) \lceil \mathbf{E}g(a) \qquad \text{Conservation of } \mathbf{E} \text{ by } g$$
$$g(a) \ddagger g(b) \qquad \text{Ex.D.6}$$

Ex.E.3. Show that a relation *g* conserves $<$, in the sense that if $a < b$, then $g(a) < g(b)$. That is, $a < b \to (g(a) < g(b))$.

We suppose that $a < b$. We have:

$a \ddagger b$		Hypothesis and def. of $<$
$a \ddagger b \to g(a) \ddagger g(b)$		Ex.E.2 above
$g(a) \ddagger g(b)$	(I)	Consequence
$\mathbf{E}g(a) = \mathbf{E}a$ and $\mathbf{E}g(b) = \mathbf{E}b$		Conservation of **E** by *g*
$\mathbf{E}a \leq \mathbf{E}b$		Hypothesis and def. of $<$
$\mathbf{E}g(a) \leq \mathbf{E}g(b)$	(II)	Consequence
$g(a) < g(b)$		(I), (II) and def. of $<$

E.2. We finally see the outline of the principal characteristics of a situation, grasped simultaneously in its being and in its appearing.

1. There is first a collection of multiples (of sets), all of which belong to the situation. This makes up the stable (and mathematically thinkable) being of every situation. We write these multiples *A*, *B*, *C*, etc.
2. Among these multiples, there is the transcendental *T*, endowed with a principally uniform structure: partial order with minimum, GI of every pair of elements, envelopes, and distributivity of the GI with regard to the envelope. Obviously, *T* can vary considerably from one situation to another, from the minimal Boolean algebra $\{M, \mu\}$, all the way to open sets of very sophisticated topological spaces.
3. Every multiple *A* of the situation can be indexed on the transcendental by a function of appearing **Id**, and the result of this indexing determines an object (*A*, **Id**). This object has atomic components of the kind $\mathbf{Id}(a, x)$, where *a* is a real element of *A*.
4. Every real element *a* of an object *A* can be assigned to its degree of existence in *A*, that is, $\mathbf{E}a$ – which is really the value of the function of appearing $\mathbf{Id}(a, a)$; and it is also localizable by an element of *T*, in the form of an atom $a \lceil p$ (restriction of *a* to *p*), which is, ontologically, defined by the element *b* such that $b(x) = a(x) \cap p$.

5 Between two objects (A, δ) and (B, γ) there can exist relations. These are functions from A to B which conserve the essential givens of appearing: intensities of existence and localizations.

6 The multiples of a situation are retroactively structured by their objectivation in appearing: compatibility, order, envelopes... Relations conserve this structuration.

A situation is finally a universe of objects and relations that makes a collection of pure multiples appear.

E.3. The question then becomes to what extent a situation, thought in its being as well as the immanent legislation of its appearing (or the transcendental constitution of objectivity), constitutes a whole. This always-singular totality, if it exists, is what combines an ontological selection (the multiples which belong to the situation) with a logical regime (the nature of the transcendental, depending on whether it is Boolean or not, does or does not provide sufficient points, is finite or infinite, etc.).

Needless to say we can only have an operational conception of this totality. It is not a question of a total 'givenness', which is meaningless, but rather of knowing which operations relating to the objects appearing in the situation can be thought *without needing to step outside of the situation*. Let's give some typical examples. Given two objects of the situation, can we think something like the sum or the total of these two objects without resorting to information that is completely extraneous to this situation? In other words, can we locate the systematic relations between one object and all the others? Are there objects that, in this regard, are privileged? And what of the question of the void, in a situation? Can we count the objects of a situation from the interior of this situation? Are there immanent resources for the classification of objects? And so on.

In studying these points, we do have a reference point, which is the pure theory of the multiple, or set theory. In this context – which is that of the ontological situation – we note that the axioms indicate which operations are possible from the interior of the universe, even though this universe cannot be presented *in* the situation, since the concept of the set of all sets is contradictory. We thus know perfectly well that there always exists the Cartesian product of two sets, the set of the parts of a set, the elevation of a set a to the power marked by a set b, etc. The fact that these operations are possible circumscribes the existence of a universe of sets.

By separating these constructions and operations from their multiple-support – and therefore, properly speaking, from the ontological situation – mathematicians and logicians have been able to propose (this is the formal substance of the most contemporary logic) a *general theory of possible universes*. In essence, we start with letters ('objects') whose signification remains indeterminate, with relations ('arrows') which are just as empty, and we define the operations in an entirely abstract way, through diagrams and algebraic calculations. In this way we produce absolutely general concepts of what the product or sum of two entities is, of what the exponentiation of one entity by another is, of what the fibred product of two relations is, etc. These minimal operations on letters and arrows provide an initial framework, which we call a category. The more operations a category admits, the richer and vaster for thought the universe it formalizes is.

Schematically, a category which admits every operation that can be performed within the basic ontological situation (set theory) will be considered a 'complete' universe, that is, a universe wherein thought moves without encountering its impossible limit. Such a category is called a *Topos*.

The complete examination of what a situation is requires, on the part of philosophy, the traversal of the theory of Topoi, since this theory gives both precision and amplitude to the concept of thinkable universe. In truth, the logic of appearing, once we advance it all the way to the dialectic of relations, requires training in categorial mathematics, just as pure ontology is unthinkable without set-theoretic mathematics.

The key point of this connection is said very simply in the following way (this result is very recent and comes from the mathematician D. Higgs, who does not utilize our philosophical lexicon…):

> THEOREM: A situation, thought in the objectivity of its appearing, namely a complete collection of objects (A, **Id**) indexed on a transcendental T and of relations between objects (set-theoretic functions preserving the degrees of existence and the localizations), is a topos, therefore a universe.

In order to demonstrate this theorem we obviously need to be more immersed in category theory than we are in this book. This is moreover an investigation that I am carrying out, under the sign of the Leibnizian question: what is a possible universe? The demonstration in effect involves verifying, on the basis of the specific definition of objects and relations, that the classical operations (sum, product, exponentiation…) are feasible without leaving the situation, and also that the logic of the situation is entirely immanent (which basically amounts to exploring the consequences of the fact that the transcendental T is one of the multiples of the situation whose transcendental it is).

Bypassing these frequently dry operational generalities, we will immediately provide some examples of the resources of every situation, thought in its generic objectivity. We will focus on the specifics of the mode of appearing of the transcendental (how that which regulates appearing can and must also appear), and on a few privileged objects, notably around the question of Zero and One.

We will begin by showing in what sense the transcendental appears as an object. The reason for this is obviously that the identity and the existence of its elements are transcendentally indexed, therefore indexed on itself.

Ex.E.4. Show that T itself, with the function of appearing (function from T to T) $\mathbf{Id}(p, q) = p \cap q$, is an object.

1 First we show that $p \cap q$ is actually a function of appearing, thus confirm the axioms of these functions, that is, **Id**.1 and **Id**.2 (cf. C1).

$\mathbf{Id}(p, q) = p \cap q = q \cap p = \mathbf{Id}(q, p)$	Verifies **Id**.1
$\mathbf{Id}(p, q) = \mathbf{Id}(q, s) = p \cap q \cap s \leq p \cap s = \mathbf{Id}(p, s)$	Verifies **Id**.2

2 Next we show that every atom of $\{T, \mathbf{Id}\}$ is real. Let $\alpha(p)$ be an atom of $\{T, \mathbf{Id}\}$. And let's posit $u = \sum\{\alpha(p) / p \in T\}$. Since $\mathbf{Id}(p, q) = p \cap q$, it follows that:

$\mathbf{Id}(u, q) = q \cap \sum\{\alpha(p) / p \in T\}$		
$\mathbf{Id}(u, q) = \sum\{\alpha(p) \cap q / p \in T\}$		Distributivity
$\alpha(p) \cap p \cap q \leq \alpha(q)$	(I)	Ax. α.1
$\alpha(p) \cap \alpha(q) \leq p \cap q$		Ax. α.2
$\alpha(p) \cap \alpha(p) \leq p \cap p$		Case where $q = p$
$\alpha(p) \leq p$		$p \cap p = p$
$\alpha(p) \cap p = \alpha(p)$	(II)	Ex.A.2
$\alpha(p) \cap p \leq \alpha(p)$		(I) and (II)

We then draw from the fact (established above) that $\mathbf{Id}(u, q) = \sum\{\alpha(p) \cap q / p \in T\}$, that $\mathbf{Id}(u, q) \leq \alpha(q)$, since $\mathbf{Id}(u, q)$, as envelope, is the least upper bound of $\alpha(p) \cap q$ (for every p), and we have just seen that $\alpha(q)$ *is* one of the upper bounds. Furthermore:

$\alpha(p) \cap \alpha(q) \leq p \cap q$ Ax. α.2
$\alpha(q) \leq q$ Case where $p = q$

In the case where $p = q$, we have $\alpha(p) \cap q = \alpha(q) \cap q$, which, by Ex.A.2 and the above result, is equivalent to $\alpha(p) \cap q = \alpha(q)$. It follows that $\mathbf{Id}(u, q)$, which must bind all the $\alpha(p) \cap q$'s, must bind $\alpha(q)$. We therefore have $\alpha(q) \leq \mathbf{Id}(u, q)$. But as we have seen that $\mathbf{Id}(u, q) \leq \alpha(q)$, antisymmetry gives us $\mathbf{Id}(u, q) = \alpha(q)$.

Consequently, the atom $\alpha(p)$ is identical to the real atom $\mathbf{Id}(u, q)$, whose ontological support is the element $u \in T$. Or, in the vocabulary of Section C, $\alpha(p) = \boldsymbol{u}$.

3 All that remains to round out the conformity of the object $(T, \mathbf{Id})$ to the axiom of materialism is to show that the real atom $\mathbf{Id}(u, q)$, or $\boldsymbol{u}$, is indeed the *only one* to be identical to the initial atom $\alpha(p)$.

By Ex.C.3, two real atoms $\boldsymbol{u}$ and $\boldsymbol{v}$ are only the same if $\mathbf{Id}(u, v) = \mathbf{E}u = \mathbf{E}v$. But if $\mathbf{Id}(u, v) = u \cap v$, we have in particular $\mathbf{E}u = u \cap u = u$, and likewise $\mathbf{E}v = v$. The condition established in Ex.C.3 then becomes $u \cap v = u = v$. Meaning that $u = v$. It is thus impossible that any real atom other than $\boldsymbol{u}$ identifies the atom $\alpha(p)$.

We will now show that the object we have just established exists in every situation (since every situation has a transcendental T, and the operation $p \cap q$ is always defined on T) has the properties of what the categoricians name a terminal object. The importance of these kinds of objects is that they fill the function of the One in a universe. They 'count for one' every object of the universe in the following sense: every object has one and only one relation with a terminal object, the relation which ensures that the object is indeed *one* object. Moreover, in the theory of Topoi, a terminal object is often referred to as the object One, and noted **1**. We will see that every situation possesses an object One, which is none other than the object $(T, \mathbf{Id}(p, q) = p \cap q)$.

Ex.E.5. There exists one and only one relation from a given object (A, δ) to the object $(T, \mathbf{Id})$.

1 Let's show first of all that a relation exists between an object (A, δ) and $(T, \mathbf{Id})$. To do this we will consider the function from A to T, called the 'existential function', which associates to every element of A its degree of existence, that is, $g(a) = \mathbf{E}a = \delta(a, a)$. This function is a relation. In effect:

— g conserves the degrees of existence:

$\mathbf{E}g(a) = \mathbf{Id}(g(a), g(a)) = \mathbf{Id}(\mathbf{E}a, \mathbf{E}a)$	Def. of $\mathbf{E}$ and of g
$\mathbf{Id}(\mathbf{E}a, \mathbf{E}a) = \mathbf{E}a \cap \mathbf{E}a = \mathbf{E}a$	Def. of $\mathbf{Id}$ and $p \cap p = p$
$\mathbf{E}g(a) = \mathbf{E}a$	Consequence

— g conserves the localizations:

An important lemma is that, for the object $(T, \mathbf{Id})$, $p \lceil q = p \cap q$. Here we take up again the convention of noting the real atom (or function) $\mathbf{Id}(a, x)$ by the name of the element a. In particular, $p \cap q$ here notes $(p \cap q)(x)$, that is, $\mathbf{Id}(p \cap q, x)$. Under these conditions, it follows that:

$$(p \lceil q)(x) = \mathbf{Id}(p, x) \cap q = p \cap x \cap q = p \cap q \cap x = \mathbf{Id}(p \cap q, x) = p \cap q$$

Moreover, $g(a \lceil p) = \mathbf{E}(a \lceil p) = \mathbf{E}a \cap p = g(a) \cap p$ (def. of g and Ex.D.3). But with the lemma, we get $g(a) \cap p = g(a) \lceil p$. And finally, $g(a \lceil p) = g(a) \lceil p$. Therefore g conserves existence and the restriction: it is a relation.

2 Now let's show that the existential function g, defined by $g(a) = \mathbf{E}a$, is the *unique* relation from (A, δ) to $(T, \mathbf{Id})$. Let's suppose that there exists an other relation f. It follows:

$\mathbf{E}f(a) = \mathbf{E}a$	Def. of f (relation)
$f(a) \cap f(a) = \mathbf{E}a$	$\mathbf{Id}(p \cap q) = p \cap q$
$f(a) = \mathbf{E}a$	$p \cap p = p$

The function f is therefore identical to the existential function g.

In order to better understand the operational extension of the universe, we will give an example of the power of the count that is established on the basis of the terminal object $(T, \mathbf{Id})$, or object One, and justify our consideration of it as the One of the situation.

Amongst the constitutive elements of an object, those which belong to it 'absolutely', namely those whose degree of existence is maximal ($\mathbf{E}x = M$), are doubtless especially remarkable. We will call *global element* of an object (A, δ) an element of this kind, that is to say a 'true' element of the object, since it appears as much as it is possible to appear.

From now on we will systematically refer to the object $(T, \mathbf{Id})$ as '**One**'. Now, we will see that there are exactly as many global elements in an object (A, δ) as there are distinct relations between **One** and this object. This means that **One** is able to count the 'absolute' elementary components of a given object. The **One** is here the determination of that which, in appearing, attests to the maximal power of appearance.

The correlation is once again made on the basis of localization. For we will show that every relation g from **One** to (A, δ) is of the kind $g(p) = a\lceil p$, where a is a fixed global element of the object (A, δ) (that is, $\mathbf{E}a = M$). So that each relation from **One** to an object counts a global element of this object. The One of the situation is the operator of the count of the 'absolute' elements of every object that appears in the situation.

The following demonstrations are good recapitulations of almost all the transcendental machinery.

Ex.E.6. Show that if a is a global element of A, $g(p) = a\lceil p$ is a relation from **One** to (A, δ).

1 The function $g(p) = a\lceil p$ conserves the existences.

$\mathbf{E}g(p) = \mathbf{E}(a\lceil p)$	Def. of g
$\mathbf{E}(a\lceil p) = \mathbf{E}a \cap p$	Ex.D.3
$\mathbf{E}a \cap p = M \cap p = p$	a is global: $\mathbf{E}a = M$
$\mathbf{E}p = p \cap p = p$	$\mathbf{Id}(p, q) = p \cap q$
$\mathbf{E}(a\lceil p) = \mathbf{E}p$	Consequence
$\mathbf{E}g(p) = \mathbf{E}p$	Def. of g

2 The function $g(p) = (a\lceil p)$ conserves the restrictions.

$g(p\lceil q) = a\lceil(p\lceil q)$	Def. of g
$p\lceil q = p \cap q$	Ex.E.4 (lemma)
$g(p\lceil p) = a\lceil(p \cap q)$	Consequence
$a\lceil(p \cap q) = (a\lceil p)\lceil q$	Ex.D.2
$g(p\lceil q) = g(p)\lceil q$	Consequence

Conserving the degrees of existence and the localizations, $g(p) = a\lceil p$ is indeed a relation from **One** to (A, δ).

Ex.E.7. Show that if g is a relation from **One** to (A, δ) there exists a global element a of (A, δ) such that $g(p) = a\lceil p$.

1 We will first show that, in (A, δ), for every relation g from **One** to (A, δ), we have the transcendental compatibility of $g(p)$ and $g(q)$ for every pair of elements of T. Namely $g(p) \ddagger g(q)$.

$\mathbf{E}g(p) \cap \mathbf{E}g(q) = \mathbf{E}p \cap \mathbf{E}q$	Conservation of existence
$\mathbf{E}p = p$ and $\mathbf{E}q = q$	$\mathbf{Id}(p, p) = p \cap p$
$\mathbf{E}g(p) \cap \mathbf{E}g(q) = p \cap q = \mathbf{Id}(p, q)$	Consequence
$\mathbf{Id}(p, q) \leq \delta[g(p), g(q)]$	Ex.E.1
$\mathbf{E}g(p) \cap \mathbf{E}g(q) \leq \delta[g(p), g(q)]$	Transitivity of $\leq$
$g(p) \ddagger g(q)$	Def. of $\ddagger$

2 Knowing that the elements $g(p)$ in A are pairwise compatible, we will use the existence of their pre-envelope (final theorem of Section C) to construct an element a of A, whose status as a global element of (A, δ) will be demonstrable. We therefore posit the existence of the pre-envelope a of the subset of A defined by the values of the function g, that is, $a = \mathbf{V}\{g(p) \,/\, p \in T\}$. Let's show that a is a global element of (A, δ).

$\mathbf{E}a = \sum\{\mathbf{E}g(p) \,/\, p \in T\} = \sum\{\mathbf{E}p \,/\, p \in T\}$	envelope a and relation g
$\mathbf{E}p = p$	$\mathbf{Id}(p, p) = p \cap p = p$
$\sum\{\mathbf{E}p \,/\, p \in T\} = \sum\{p \,/\, p \in T\} = M$	Consequence, and maximum M
$\mathbf{E}a = M$, therefore a is a global element	Consequence

3 Next we will demonstrate in stages that for every $q \in T$, $g(q) = a \lceil q$, where a is the global element of (A, δ) constructed above. This will finally establish what we announced in the introduction of the last two exercises: every relation from **One** to (A, δ) is indeed of the form $g(q) = a \lceil q$, where a is a global element of (A, δ).

4 First let's investigate exactly what $a \lceil q$ means.

$a \lceil q = \mathbf{V}\{g(p) \,/\, p \in T\} \lceil q$	Def. of a
$\mathbf{V}\{g(p) \,/\, p \in T\} \lceil q = \mathbf{V}\{g(p) \lceil q \,/\, p \in T\}$	Ex.D.12
$\mathbf{V}\{g(p) \lceil q \,/\, p \in T\} = \mathbf{V}\{g(p \lceil q) \,/\, p \in T\}$	g is a relation
$p \lceil q = p \cap q$	Ex.5.4 (lemma)
$a \lceil q = \mathbf{V}\{g(p \cap q) \,/\, p \in T\}$	Consequence

5 A lemma, which is very interesting in itself, is that the order-relation induced by the transcendental objectivity within **One** (considered as an object), that is, $<$, coincides with the essential relation that is the transcendental structure of T, the relation $\leq$. Let's show this.

We have $p < q$ if and only if $\mathbf{E}p \leq \mathbf{E}q$ and $p \ddagger q$ (Ex.D.5). However, in the object $(T, \mathbf{Id})$, or **One**, we know that $\mathbf{E}p = p$ and $\mathbf{E}q = q$ (cf. demonstration of Ex.E.6). Consequently, $\mathbf{E}p \leq \mathbf{E}q$ is exactly the same thing as $p \leq q$. Moreover, we have by definition $p \ddagger q$ if and only if $\mathbf{E}p \cap \mathbf{E}q = \mathbf{Id}(p, q)$. However, this is always true in **One**, since equality amounts to $p \cap q = p \cap q$ (let's note in passing that in this object, which serves as the One for the entire situation, all the elements are pairwise compatible). Finally, the relation $p < q$ is exactly identical to $p \leq q$.

6 We know from Ex.E.3 that a relation conserves the order-relation $<$. Taking into account the above paragraph, it follows:

$p \leq q \rightarrow g(p) < g(q)$	g conserves order
$p \cap q \leq q$	Def. of $\cap$
$g(p \cap q) < g(q)$	Consequence

But we have shown that $a \lceil q$ was the pre-envelope of $\{g(p \cap q) \, / \, p \in T\}$. The final theorem of Section C tells us that $a \lceil q$ is also the *envelope* of this subset of A for the order-relation $<$. It is therefore the least upper bound of all the $g(p \cap q)$'s. And as we have just shown that $g(q)$ is one of these upper bounds, we must conclude that:

$$a \lceil q < g(q) \qquad \text{(I)}$$

On the other hand, in the case where $p = q$, it is clear that $g(p \cap q) = g(q)$. With regard to its function as upper bound of $g(p \cap q)$ for all the possible values of p, we necessarily have:

$$g(q) < a \lceil q \qquad \text{(II)}$$

The comparison of (I) and (II) gives, by antisymmetry, $g(q) = a \lceil q$. This proves that the relation g from **One** to (A, δ) does indeed correspond to a global element a such that such that $g(p) = a \lceil p$.

There is therefore (categorial) identity between the relations from **One** to the object (A, δ) and the global elements of (A, δ), namely those whose degree of existence $\mathbf{E}a$ is maximal. We say that, in the object (A, δ), **One** serves as a *count-for-one* of the global elements. This is a second major reason for naming this object **One**.

E.4. After the construction of a terminal object (there is a unique function from every object toward it), namely **One**, we turn toward the 'initial' object (there is a unique function from it toward every object). This object is also named **Zero**, and we will effectively see that, in its construction, the role of the minimum μ and that of a kind of inexistent element are determinants.

For every object (A, δ), the partition function $\alpha(x) = \mu$ is an atom, as can be immediately confirmed by the fact that $\alpha(x) \cap \alpha(y) = \mu$ and $\alpha(x) \cap \delta(x, y) = \mu$, which verifies the axioms α.1 and α.2 of the atomic components of objects.

By virtue of the axiom of materialism, which requires that every atom is real, there exists an element of A which identifies this atom. If $\varnothing_A$ is this element, we have, for every $x \in A$, $\delta(\varnothing_A, x) = \mu$. In particular, $\mathbf{E}\varnothing_A = \delta(\varnothing_A, \varnothing_A) = \mu$. The degree of existence of $\varnothing_A$ is minimal, or transcendentally nil.

It is worthwhile dwelling on the signification of this spectral element. Its 'existence' is guaranteed in every object, since the minimum μ exists in every transcendental, and so too does, for every multiple A of the situation (thought ontologically), the atom $\varnothing_A$. However, since its degree of existence is nil and its identity to every other element of A is also nil, we can say that it is *the proper inexistent of* A. The functions of such a local inexistent (as distinct from the empty set, which is being as non-being, and which is a global operator) are considerable, and philosophical meditation on this subject is very important.

We will now construct the schema of the Zero of every situation.

Let's consider the set $\{\mu\}$, the singleton of μ. For this set, the function 0 defined by $0(\mu, \mu) = \mu$ is a function of appearing (the verification of the axioms **Id**.1 and **Id**.2 is trivial). We will call the object $(\{\mu\}, 0)$ **Zero**.

That every atom of the object **Zero** is real, and that it is therefore a real object (for the materialist investigation) is easily demonstrated. Let $\alpha(\mu)$ be a presumed atom of **Zero**. The atomic axiom α.2 demands that we have: $\alpha(\mu) \cap \alpha(\mu) \leq 0(\mu, \mu) = \mu$. There therefore exists only a single atom, defined by $\alpha(\mu) = \mu$, which is also real, since it is identical to the atom μ, in the sense where $\mu(\mu) = (\mu, \mu) = \mu$. We will note in passing that, in the object **Zero**, the 'ghostly' element $\varnothing_A$ is none other than μ, thus μ, **Zero** has no element other than $\varnothing_A$. The degree of existence of the unique element of **Zero** is therefore nil. **Zero** is that object which makes One of an inexistent.

Ex.E.8. Show that there exists one and only one relation from **Zero** to an object (A, δ).

1 There exists a relation:
Let $\varnothing_A$ be the element of A which realizes the atom $\alpha(x) = \mu$. Let's posit that $g(\mu) = \varnothing_A$. We will show that this is a relation from **Zero** to (A, δ). In effect:

$\mathbf{E}(g(\mu)) = \mathbf{E}\varnothing_A = \mu = 0(\mu, \mu) = \mathbf{E}\mu.$	g conserves $\mathbf{E}$
$\mu \lceil p = 0(\mu, \mu) \cap p = \mu \cap p = \mu$	Def. of $\lceil$ and $p \cap \mu = \mu$
$g(\mu \cap p) = g(\mu) = \varnothing_A$	Consequence
$(\varnothing_A \lceil p)(x) = \delta(\varnothing_A, x) \cap p = \mu \cap p = \mu = \varnothing_A(x)$	Various obvious points
$\varnothing_A = \varnothing_A \lceil p = g(\mu) \lceil p$	Consequence
$g(\mu \lceil p) = g(\mu) \lceil p$	g conserves $\lceil$

2 There exists only one relation:
Given a relation g from **Zero** to (A, δ), we will show this is the same as the one above (that is, $g(\mu) = \varnothing_A$).

Let's suppose that $g(\mu) = t$. If g is a relation, there is a conservation of **E**, and we have $\mathbf{E}g(\mu) = \mathbf{E}t = \mathbf{E}\mu = \mu$. It then follows:

$\delta(t, t) = \mu$	$\mathbf{E}t = \mu$
$\delta(t, t) \cap \delta(t, x) = \mu$	$p \cap \mu = \mu$
$\delta(t, x) = \mu$ for every x	Ax. **Id**.2
$t = \varnothing_A$	Def. of $\varnothing_A$

The relation g is thus the relation $g(\mu) = \varnothing_A$, and there is no other.

Ex.E.9. If a relation from **One** to **Zero** exists, the transcendental discriminates nothing, and its operation is void.

We have seen (Ex.E.6.) that if a relation from **One** to **Zero** exists, then there must exist a global element of **Zero**, that is, $a \in \{\mu\}$ with $\mathbf{E}a = M$. But the only (real) element of $\{\mu\}$ is μ. And in the object **Zero**, $\mathbf{E}\mu = \mu$. We must therefore finally have, in the transcendental T, $\mu = M$. But as we have $\mu \leq p \leq M$ for every supposed element p of T, if $\mu = M$ then T is obviously reduced to a single element, or $T = \{\mu\}$, with $\mu \leq \mu$ as its order. This transcendental obviously has no power of differentiation, and therefore cannot operate on anything.

We see that, for every situation, the transcendental only operates if it excludes from the field of relations everything which claims to count the zero as one. We will say that a situation is consistent, with regard to its appearing, only if its object **Zero** – which always exists – remains subtracted from the relational operations of the count which have their origin in the **One**.

We can on this point adopt another language. Let's say that an object is void if it has no global element. Thus, if no relation from **One** to this object exists (cf. exercises E.6. and E.7.), we will then say that a condition of transcendental consistency for appearing is that **Zero** is void.

In the ontological situation, which is nothing other than the mathematics of sets, there is the following essential property: **Zero** and void are one and the same thing. In other words, the initial object, namely the **Zero**, which is none other than the empty set $\varnothing$, is the unique set that has no global element, therefore the unique set which is void in the sense indicated above. In fact, in the ontological situation, every belonging is global, or, there are only global elements. Because, given a multiple A, either x in it is a global element, that is, $x \in A$, or it is not an element at all, that is, $x \notin A$. There remains a crucial link between this characteristic and the fact that the 'transcendental' of the ontological situation is none other than the minimal transcendental $\{\mu, M\}$.

This situation is, however, exceptional. In a given situation, the transcendental can be absolutely different from the classical minimal transcendental $\{\mu, M\}$, and the functions of appearing do not have the rigidity of the function $\mathbf{Id}(x, y) = M$ if $x = y$, and $\mathbf{Id}(x, y) = \mu$ if $x \neq y$. Appearing can envelop multiple nuances, and the degrees of existence can be very differentiated, or even, as is most often the case, infinite in number. It follows that there is no requirement that **Zero** be the only empty object. To the contrary, one can find numerous objects devoid of any global element, and which are moreover in no way identical to **Zero**.

The question of whether or not, in a situation, **Zero**, which is always void, is the only empty object, is a fundamental onto-logical question, the question of *the uniqueness of the void*. Situations have very different characteristics depending on whether or not, like the ontological situation, they admit only one empty object, or contrarily – and on this point we are far removed from the thinking of being (which is always underlying) – they authorize, in the play of appearing, numerous different empty objects.

On this subject there is an absolutely remarkable theorem worth mentioning. Let's say that a situation is *well-differentiated* if, given two different relations from an object (A, δ) to an object (B, γ) of the situation, these relations differ on at least one *global* element of A.

In other words: let f be a relation from (A, δ) to (B, γ), and let g be another relation between the same objects. Both relations are therefore presumed to differ intrinsically, and not by the objects they link. This assumes that for at least one element of the set A, we have $f(a) \neq g(a)$.

Otherwise both relational functions would be indiscernible. We will say that the situation is well-differentiated if, whenever such a configuration presents itself, there exists at least one global element of A, say a – thus an element such that $\mathbf{E}a = M$ – for which we effectively have $f(a) \neq g(a)$.

This means that a situation is well-differentiated if, in the relational universe which constitutes its appearing, two different relations concerning the same objects see their difference attested in an 'absolute' point of the object which shares the relation, a point whose existence in this object is absolute. And as a relation conserves the existences, the element that corresponds by the relation to the certain element will itself be a certain element (we will have $\mathbf{E}f(a) = M$). In this sense, the difference between two relations is attested *locally* by an element whose appearing has the maximal intensity, whose existence, in the two objects connected to the situation, is absolutely certain. If this is not the case, the difference between the relations remains global, or elusive, due to the fact that no certain support is given to it locally.

The theorem that we announced earlier (but that we do not demonstrate here) is then the following: if a situation is well-differentiated then, in this situation, **Zero** is the only empty object. Or: if the difference between two relations is always attested in an existentially certain point, then the void is unique.

This theorem is onto-logical in the strong sense. Because it links a property of appearing, or of existence (its maximal value) to an ontological characteristic: the uniqueness of the void.

APPENDIX: ON THREE DIFFERENT CONCEPTS OF IDENTITY BETWEEN TWO MULTIPLES OR TWO BEINGS

The key to transcendental theory is, as we have seen, the theory of identity functions **Id**, that is, the indexation of the degree of identity between two elements of a set *A*, presented in a situation *S*, on the transcendental of this situation (mathematically: on a complete Heyting algebra). We then obtain the possible nuances of identity in appearing – or of the local appearing of identity – nuances which are distributed between maximal or absolute identity (M), leading to the certainty of the Same, and minimal identity (μ), or absolute difference, leading to the certainty of the Other. The intermediary values distribute the intermediary identities, or degrees of resemblance, that are characteristic of appearing.

Regarding identity, or equality with itself, indexation fixes the degree of existence (with respect to the ontological referential *A* and the transcendental *T* of the situation where *A* is presented) between absolutely certain existence ($\mathbf{E}x = M$) and guaranteed inexistence ($\mathbf{E}x = \mu$).

A question however remains. The elements *x* we are speaking of are supposed to 'exist' in *A* in the ontological sense of the term, even if their degree of appearing is nil. We have not yet said anything as to the *treatment of the inexistent as such*, of the ontological inexistent. In particular, what can we think and say about the identity of inexistents? Are two beings (two multiples in the ontological sense) 'outside of existence' required to be identical? Or are they subtracted from every identitarian comparison?

This point effectively divides the very notion of equality, once, at play in appearing, it is compared not only with multiple-being but with its existence and its degrees.

We have seen (Ex.C.1.) that a fundamental property of the functions of appearing is written: $\mathbf{Id}(x, y) \leq \mathbf{E}x$, a property which indicates that one cannot be more identical to another than one is to oneself, or that one can only identify with another to the extent that one exists, and no more.

But by the property of dependence established in Ex.A.6, we can equally write: $[\mathbf{Id}(x, y) \Rightarrow \mathbf{E}x] = M$. It is also certain that it will be in the situation (maximal degree M) that the degree of identity between x and y entails, or envelops, the degree of existence of x (or of y).

In other words: it is absolutely *true* that the evaluation of identity **Id** between x and y requires taking into account the existence of x (or equally of y) and disallowing the identitarian evaluation of beings that are subtracted from all existence. The complete logic of appearing, however, supposes that we can also address identity, or a certain identity, without any existential limit, according to the ontological principle, dating back to Plato, of a certain being of non-being. The means of this extension is simple: we will assume that *two inexistents are identical*. And that if we consider them as existents, then we return to the evaluation by the functions **Id**. This supposes the definition of another identity, dependent, of course, on **Id**, but which suspends the prevalence of existence, replacing it with the status of a condition (whose negation is admissible) and therefore includes the thesis of the identity of all the inexistents. We will call this new function *equivalence* (in appearing).

DEFINITION: Given an object $(A, \mathbf{Id})$, we call *equivalence* of x and y, and write this $\varepsilon(x, y)$, the following function, which to every pair of elements x and y of A associates a value in the transcendental T:

$$\varepsilon(x, y) = (\mathbf{E}x \cap \mathbf{E}y) \Rightarrow \mathbf{Id}(x, y)$$

Let's comment on this formula. The value of equivalence of x and y is that of their identity $\mathbf{Id}(x, y)$, under condition of the values of existence of x and y. Let's illustrate this idea with some examples.

a If $\mathbf{E}x = \mathbf{E}y = \mu$, or if x and y absolutely inexist for A, we have $(\mathbf{E}x \cup \mathbf{E}y) = \mu$. But then, knowing that $(\mu \Rightarrow q) = M$ regardless of whatever q may be (this is the principle *ex falso sequitur quodlibet*, cf. B.2), we get: $[(\mathbf{E}x \cup \mathbf{E}y) \Rightarrow \mathbf{Id}(x, y)] = M$. And therefore $\varepsilon(x, y) = M$. Which means: two inexistents are absolutely equivalent.

b With regard to the difference of $\mathbf{E}x$, which evaluates $\mathbf{Id}(x, x)$, thus equality to oneself, which is a function of existence that can vary from μ to M, the self-equivalence of x is always absolutely true. In other words, $\varepsilon(x, x) = M$. In fact, $\varepsilon(x, x) = (\mathbf{E}x \cup \mathbf{E}x) \Rightarrow \mathbf{Id}(x, x)$. But, by definition, $\mathbf{Id}(x, x) = \mathbf{E}x$. And $\mathbf{E}x \cup \mathbf{E}x = \mathbf{E}x$. We finally get: $\varepsilon(x, x) = [\mathbf{E}x \Rightarrow \mathbf{E}x] = M$.

c The value of equivalence ε will logically never be inferior to that of identity **Id**. Or: as identity is more exacting (since it does not extend to the inexistents), it always has a degree of truth less than or equal to that of equivalence, which is more permissive (two inexistents are equivalent). We therefore must expect: $\mathbf{Id}(x, y) \leq \varepsilon(x, y)$.
And this is precisely what we have.

Ex.App.1. Show that $\mathbf{Id}(x, y) \leq \varepsilon(x, y)$

$\varepsilon(x, y) = (\mathbf{E}x \cap \mathbf{E}y) \Rightarrow \mathbf{Id}(x, y)$	Def. of ε
$q \cap (p \Rightarrow q) = q$	Ex.A.8.
$\mathbf{Id}(x, y) \cap \varepsilon(x, y) = \mathbf{Id}(x, y)$	Application
$\mathbf{Id}(x, y) \leq \varepsilon(x, y)$	Ex.A.2.

d If we now *add* the degree of existence of x and y to their equivalence, the total value must indeed be that of their identity. We logically expect: $\mathbf{Id}(x, y) = \varepsilon(x, y) \cap \mathbf{E}x \cap \mathbf{E}y$.

This statement expresses that identity in appearing is the *conjunction of equivalence and existence.*

Ex.App.2. Show that $\mathbf{Id}(x, y) = \varepsilon(x, y) \cap \mathbf{E}x \cap \mathbf{E}y$

$\mathbf{Id}(x, y) \leq \varepsilon(x, y)$		Above exercise
$\mathbf{Id}(x, y) \leq \mathbf{E}x$ and $\mathbf{Id}(x, y) \leq \mathbf{E}y$		Ex.C.1.
$\mathbf{Id}(x, y) \leq \varepsilon(x, y) \cap \mathbf{E}x \cap \mathbf{E}y$	(I)	Def. of GI
$p \cap (p \Rightarrow q) \leq q$		Ex.A.7.
$[\mathbf{E}x \cup \mathbf{E}y] \cap [(\mathbf{E}x \cup \mathbf{E}y) \Rightarrow \mathbf{Id}(x, y)] \leq \mathbf{Id}(x, y)$		Application
$[\mathbf{E}x \cup \mathbf{E}y] \cap \varepsilon(x, y) \leq \mathbf{Id}(x, y)$		Def. of ε
$\mathbf{E}x \cap \mathbf{E}y \leq \mathbf{E}x \cup \mathbf{E}y$		Evidence

$[\mathbf{E}x \cap \mathbf{E}y] \cap \varepsilon(x, y) \leq [\mathbf{E}x \cup \mathbf{E}y] \cap \varepsilon(x, y)$		Consequence
$[\mathbf{E}x \cap \mathbf{E}y] \cap \varepsilon(x, y) \leq \mathbf{Id}(x, y)$	(II)	Consequence
$\mathbf{Id}(x, y) = \mathbf{E}x \cap \mathbf{E}y \cap \varepsilon(x, y)$		Antisymmetry (I), (II)

We actually have, between being and appearing, three concepts of identity (or three categories of the Same).

- The *ontological* concept, $x = y$ or the identity of two pure multiples, which is regulated by the axiom of extensionality: x and y are identical if they have the same elements.
- The logical concept of *equivalence*, $\varepsilon(x, y)$, which treats all the inexistents as identical and systematically gives a maximal value to the equivalence of an element to itself.
- The concept bound up with *effective appearing*, or transcendental identity, which evaluates the identity of two beings under condition of a value of existence which is supposed to always be superior to that of its identity.

The first concept is Boolean. Because, as we have seen, the belonging or non-belonging of x to A, and the identity of A, are regulated by a function on the minimal algebra $\{\mu, M\}$.

The other two concepts suppose a space for evaluation and localization which depends on being-there, or on the there of being, on appearing as logic, and therefore on transcendental operations which, in general, are not Boolean, although they may be in certain situations.

TRANSLATOR'S ENDNOTES

1 Between the publications of *L'Être et l'événement* in 1988 and *Logiques des mondes* in 2006 Badiou produced one novel (*Calme bloc ici-bas*), four plays (*La Tétralogie d'Ahmed*), and no fewer than 15 major philosophical works, one devoted entirely to mathematics (*Number and Numbers*, trans. Robin Mackay, Cambridge: Polity Press, 2008) and at least three of which quickly became bestsellers, namely, *Ethics: An Essay on the Understanding of Evil* (trans. Peter Hallward, London: Verso, 2002), *Saint Paul: The Foundation of Universalism* (trans. Ray Brassier, Stanford, California: Stanford University Press, 2003), and *Deleuze: The Clamor of Being* (trans. Louise Burchill, Minneapolis: University of Minnesota Press, 2000).

2 See Alain Badiou, *Logics of Worlds: Being and Event, 2*, (trans. Alberto Toscano, London: Continuum, 2009), 37–8: 'Just as being *qua* being is thought by mathematics ... so appearing, or being-there-in-a-world, is thought by logic. Or more precisely: "logic" and "consistency of appearing" are one and the same thing.'

3 Alain Badiou, *Topos, ou Logiques de l'onto-logique: Une introduction pour philosophes, tome 1* (unpublished typescript), 1/13 (all subsequent references to this and to *L'Être-là: Mathématique du transcendental* will include the page number of the original French typescripts followed by the page number in the present English translation, separated by a slash). See also: 'Set theory and category theory therefore offer distinct paths for all the decisive questions regarding the thinking of being (acts of thought, forms of immanence, identity and difference, schools of logic, admissible rationally, relation of experience to existence, infinity, unity or plurality of universes, etc.). This is to say that they *lay out different conditions for philosophy*' (5/16).

4 Philosophy's unique objective, according to Badiou, is 'the thinking of thought' (*Deleuze*, 21), that is, thinking the compossibility of the various (artistic, political, amorous, scientific) truths.

5 Alain Badiou, *Briefings on Existence: A Short Treatise on Transitory Ontology* (trans Norman Madarasz Albany, NY: SUNY Press, 2006), x.

6 The work with which *Topos* bears the most immediate comparison is in fact Robert Goldblatt's canonical text *Topoi: The Categorial Analysis of Logic* (New York: Dover, 1984), a work which Badiou acknowledges as his own introduction to category theory (see: Badiou, *Logics of Worlds*, 538; Badiou, *L'Être-là*, 7/169) and that also concerns itself foremost with the differences between set and category theory, and which itself displays an admirable philosophical bent.

7 In addition to the texts he singles out in the present book – namely, the canonical works of Goldblatt, Fourman, Scott and Higgs (see notes 40 and 41) – Badiou lists as his principal references Francis Borceux's three-volume collection *Handbook of Categorical Algebra* (Cambridge: Cambridge University Press, 1994), Oswald Wyler's *Lecture Notes on Topoi and Quasitopoi* (Singapore: World Scientific, 1991), and J. L. Bell's *Toposes and Local Set Theories* (Oxford: Oxford University Press, 1991), together with a number of foundational articles. It should be pointed out that these additional texts are fundamental for Badiou specifically with regard to the concept of complete Heyting algebras, which serve as the mathematical expression of what Badiou calls the 'transcendental' of a world (and which is one of the principal concepts elaborated here in Book II). See: Badiou, *Logics of Worlds*, 537–9.

8 Badiou argues in *The Century* (trans. Alberto Toscano, Cambridge: Polity, 2007) that the twentieth century was marked by a pronounced passion for the real, which is equally a passion for formalization: 'Formalization is basically the great unifying power behind all the century's undertakings – from mathematics (formal logics) to politics (the Party as *a priori* form of any collective action), by way of art, be it prose (Joyce and the odyssey of forms), painting (Picasso, the inventor of a suitable formalization in the face of every occurrence of the visible) or music (the polyvalent formal construction of Alban Berg's *Wozzeck*)' (160).

9 Alain Badiou, *The Concept of Model: An Introduction to the Materialist Epistemology of Mathematics* (ed. and trans. Zachary Luke Fraser and Tzuchien Tho, Melbourne: re.press, 2007).

10 It should also be pointed out that *Mathematics of the Transcendental* does not in fact mark Badiou's initial encounter with category theory. To the contrary, while he is yet to yet explore the branch in any detail, references to 'the theory of categories' in Badiou's work appear as early as his 1967 article on 'The (Re)commencement of Dialectical

Materialism' (where he declares category theory to be 'perhaps the most significant epistemological event of these last years, due to the radical effort of abstraction to which it bears witness)', Alain Badiou, 'The (Re) commencement of Dialectical Materialism', in *The Adventure of French Philosophy* (ed. and trans. Bruno Bosteels London: Verso, 2012), 166.

11 The Cercle d'Épistémologie (Epistemology Circle) were a loose collective of students at the École normale supérieure that counted among its members Jacques-Alain Miller, Jean-Claude Milner, François Regnault, Alain Grosrichard, and Jacques Bouveresse. The *Cahiers pour l'Analyse* was the influential journal they published between 1966–9. Key texts from the journal as well as interviews with Cercle members have recently been published as *Concept and Form, Volume 1: Selections from the Cahiers pour l'Analyse*, Peter Hallward and Knox Peden (eds) (London: Verso, 2012) and *Concept and Form, Volume 2: Selections from the Cahiers pour l'Analyse*, Peter Hallward & Knox Peden (eds) (London: Verso, 2012).

12 See: Alain Badiou, 'Infinitesimal Subversion', *Concept and Form, Volume 1*, 187–207; Alain Badiou, 'Mark and Lack: On Zero', *Concept and Form, Volume 1*, 159–95. Discussing his time with the Cercle d'Épistémologie, Badiou concedes that he was 'at the extreme point of a strict formalism. I pushed much further than my friends the detailed study of the recent developments of mathematical logic, notably the sectors in full effervescence comprising set theory (Cohen's theorem) or the new non-standard theory of numbers', Alain Badiou, 'Theory from Structure to Subject: An Interview with Alain Badiou', *Concept and Form, Volume 2*, 280.

13 Alain Badiou, *Theory of the Subject*, trans. Bruno Bosteels (London: Continuum, 2009).

14 See Jacques Lacan, *The Seminar of Jacques Lacan Book XX: On Feminine Sexuality, the Limits of Love and Knowledge, 1972–1973* (trans. Bruce Fink New York: Norton, 1998), 119: 'Mathematical formalization is our goal, out ideal.'

15 Alain Badiou, *Being and Event* (trans. Oliver Feltham, London: Continuum, 2005).

16 See Badiou, *Briefings on Existence*, 40 (trans. modified): 'Given that no immanent limit originating in the One determines multiplicity as such, there is no first principle of finitude. The multiple can therefore be considered in-finite. Or again: infinity is another name for multiplicity as such. And as no principle binds infinity to the One, we must maintain that there is an infinity of infinities, an infinite dissemination of infinite multiplicities.'

17 ZFC stands for Zermelo-Fraenkel with the axiom of choice, being the branch of mathematics that is by far the most frequently employed foundational system of mathematics. The axiom of choice states that, given a collection of non-empty sets *A* which have no element in common, there exists a set *C* which has one and only one element in common with every set *a* of *A*. 'In other words', Badiou explains in *Being and Event*, 'one can "choose" an element from each of the multiples which make up a multiple, and one can 'gather together' these chosen elements' (224). While the axiom is controversial (many mathematicians remain opposed to it), it is crucial to Badiou's conception of a truth-procedure (in particular his formulation of subjective intervention). It is also worth noting that Badiou doesn't hold ZFC to be the only *possible* discourse on being (and by extension the only possible ontology), but rather the best system available: that ontology is *historically accomplished* as a mathematics of multiplicities does not demand that set theory alone provides this structure. It is always conceivable something better will come along, one obvious contender being category theory.

18 Other pertinent aspects of ZFC include its axiomatic (i.e. non-derivative, non-deductive, non-intuitive) nature; the fact that it is entirely expressed in first-order logic and that all of its concepts are defined intrinsically; that it admits of only a single primitive relation (belonging); that it possesses only one existential axiom (the axiom of the void); that it is non-descriptive and non-empirical; and that it axiomatically constructs its universe from the void alone (through the axioms of the void and the power set).

19 Of the many exegeses available today we direct the reader in particular to Peter Hallward's still-indispensible *Badiou: A Subject to Truth* (Minneapolis: University of Minnesota Press, 2003) and the collection *Alain Badiou: Key Concepts*, A. J. Bartlett and Justin Clemens (eds) (Durham: Acumen, 2010). Of course it has to be said that Badiou is his own best exegete and *Being and Event* remains far and away the best introduction to the role of set theory in his philosophy.

20 See Badiou, *Briefings on Existence*, 34 (trans. modified): 'We can define metaphysics as follows: the *enframing* of being by the one. Its most appropriate synthetic maxim is that of Leibniz, which establishes the reciprocity of being and the one as a norm: "what is not *a* being is not a *being*".' Badiou begins the first meditation of *Being and Event* in a similar manner: 'Since its Parmenidean organization, ontology has built the portico of its ruined temple out of the following experience: what *presents* itself is essentially multiple; *what* presents itself is essentially one' (23). Interestingly, prior to his equation of ontology with mathematics (and consequent ability to align being with pure multiplicity) in *Being and Event*, Badiou held that philosophies of the multiple were in fact covert metaphysics of the One. Thinking in particular of Deleuze – whom he later recognized as his principal interlocutor in developing his own theory of the multiple (cf. *Deleuze*, 3) – and his philosophy of multiplicities, Badiou argues in *Theory of the Subject* that such a multiple is in truth 'never more than a semblance since positing the multiple amounts to presupposing the One as substance and excluding the Two from it. The ontology of the multiple is a veiled metaphysics' (22). On Deleuze's supposed clandestine embrace of the One, see also Badiou's scathing 'The Fascism of the Potato' in *The Adventure of French Philosophy*, ed. and trans. Bruno Bosteels (London: Verso, 2012), 191–201. Of course Badiou expands considerably on this thesis in *Deleuze: The Clamor of Being*.

21 Badiou, *Briefings on Existence*, xi.

22 Badiou, *Being and Event*, xi.

23 Badiou, *Briefings on Existence*, ix–x.

24 Again, these doubts (which are of course finally 'answered' in *Logics of Worlds*) are already well documented. Badiou himself addresses them almost as a matter of course in the prefaces to the English language editions of his later works (see in particular the introductions to *Ethics* and *Briefings on Existence*). The text that deals most immediately with these reformulations is, for obvious reasons, his 'transitional' work, *Briefings on Existence*. One should also consult the collections *Theoretical Writings*, (ed. and trans. Ray Brassier and Alberto Toscano, London: Continuum, 2004) and *Think Again: Alain Badiou and the Future of Philosophy*, ed. Peter Hallward (London: Continuum, 2004), as well as Hallward's (who, after Jean-Toussaint Desanti, was one of the first to criticize Badiou's non-relational ontology) *Badiou: A Subject to Truth*.

25 As Badiou puts it in *Logics of Worlds*, 'in *Being and Event*, I did not interrogate the fact that every givenness of being takes the form of a situation. We could say that the realization of being-qua-being as being-there-in-a-world is granted as though it were a property of being itself. But this (Hegelian) thesis can only be defended if one attributes to being the *telos* of its appearing. Only if, basically, one agrees that it is of the essence of being to bring about worlds in which its truth manifests itself' (36).

26 In ZFC the axiom of extensionality prescribes that two sets are absolutely the same if they have exactly the same elements. Accordingly, for two sets to be absolutely different it suffices that one of the sets has a single element that the other does not possess.

27 As Badiou summarizes here, 'ontologically, one multiple cannot be "more or less" different to another multiple. A multiple is only identical to itself, and it is a law of being *qua* being (axiom of extensionality) that the least local difference, bearing for example on one single element amongst an infinity of others, entails an absolutely global difference. It will not be the same for appearing. It is clear that, in a situation, multiples can be more or less different, similar, proximate, etc. We must therefore admit that what governs appearing is not the ontological composition of a particular being (a multiple), but the relational evaluations that determine the situation and localize this being within it. Unlike the legislation of the pure multiple, these evaluations do not always equate local and global difference. They are not ontological. This is why we will call logic the law of the network of relations that determines appearing in the situation of multiple-being. Every situation possesses its own logic, which legislates on appearing, or the "there" of being-there', Badiou, *L'Être-là*, 5/167.

28 Desanti taught philosophy at both the École normale supérieure and at the Sorbonne (where he succeeded George Canguilhem). His most celebrated work, adapted from his doctoral thesis, is *Les Idéalités mathématiques: Recherches épistémologiques sur le développement de la théorie des fonctions de variables réelles* (Paris, Seuil, 1968).

29 Jean-Toussaint Desanti, 'Quelques Remarques à propos de l'ontologie intrinsèque d'Alain Badiou', *Les Temps modernes*, vol. 526 (May 1990), 61–71. Translated by Ray Brassier as 'Some Remarks on the Intrinsic Ontology of Alain Badiou' in *Think Again: Alain Badiou and the Future of Philosophy*, ed. Peter Hallward (London: Continuum, 2004), 59–66.

30 Desanti, 'The Intrinsic Ontology of Alain Badiou', 66.

31 Desanti, 'The Intrinsic Ontology of Alain Badiou', 66.

32 Desanti, 'The Intrinsic Ontology of Alain Badiou', 59.

33 Desanti, 'The Intrinsic Ontology of Alain Badiou', 59.

34 Badiou's is moreover a minimal intrinsic ontology insofar as its thought hinges on the gap between the bare minimum effect of structure and what is radically unstructured, namely, between the operation of the count and the abyss of pure multiplicity, between structural consistency and the formlessness of the void, between the situation and its in-consistent underside.

35 Desanti, 'The Intrinsic Ontology of Alain Badiou', 59.

36 Badiou, *L'Être-là*, 1/165.

37 On the development of set theory in response to problems in Analysis (in particular, the problem of infinite quantities), see Alain Badiou, *Second*

Manifesto for Philosophy (trans. Louise Burchill, Cambridge: Polity Press, 2011), 111–15. See also: Badiou, *Number and Numbers*, 52–8; and Badiou, 'Infinitesimal Subversion', 187–207.

38 Badiou's neologism combines the French word for 'in' (*dans*) and the German word for being, (*sein*), in a way that both interrupts and extends the Heideggarian usage of the term Dasein which denotes, for Heidegger, the Being in(-the-world) of which the human entity is capable. Cf. Martin Heidegger, *Being and Time* (trans. John Macquarrie and Edward Robinson, New York: HarperCollins Publishers, 1962), 27.

39 See Jacques Lacan, *The Seminar of Jacques Lacan Book XI: The Four Fundamental Concepts of Psychoanalysis* (trans. Alan Sheridan, New York: Norton, 1998), 212–13.

40 Robert Goldblatt, *Topoi: The Categorial Analysis of Logic* (New York: Dover, 1984).

41 See: Michael P. Fourman and Dana S. Scott, 'Sheaves and Logic', in *Applications of Sheaf Theory to Algebra*, M. P. Fourman, C. J Mulvey, and D. S. Scott (eds) (Berlin: Springer-Verlag, 1979), 302–401; and Denis Higgs, *A Category Approach to Boolean-Valued Set Theory* (Lecture Notes, University of Waterloo, 1973).

INDEX

www.ingramcontent.com/pod-product-compliance
Ingram Content Group UK Ltd.
Pitfield, Milton Keynes, MK11 3LW, UK
UKHW020717280225
455688UK00012B/401